Tactical Approaches to Technical Communication

SUNY series, Studies in Technical Communication

Miles A. Kimball, Derek G. Ross, and Hilary A. Sarat-St. Peter, editors

Tactical Approaches to Technical Communication

Reimagining Institutions, Transforming Society

Edited by

HAYLEY MCCULLOUGH,
HILARY A. SARAT-ST. PETER,
and MILES A. KIMBALL

SUNY
PRESS

Published by State University of New York Press, Albany

Printed in the United States of America

EU GPSR Authorised Representative:
Logos Europe, 9 rue Nicolas Poussin, 17000, La Rochelle, France
contact@logoseurope.eu

For information, contact State University of New York Press, Albany, NY
www.sunypress.edu

Library of Congress Cataloging-in-Publication Data

Name: McCullough, Hayley, 1992– editor. | Sarat-St. Peter, Hilary A., editor. | Kimball, Miles A., editor.
Title: Tactical approaches to technical communication : reimagining institutions, transforming society / edited by Hayley McCullough, Hilary A. Sarat-St. Peter, and Miles A. Kimball.
Description: Albany : State University of New York Press, [2025] | Series: SUNY series, studies in technical communication | Includes bibliographical references and index.
Identifiers: LCCN 2024042771 | ISBN 9798855802061 (hardcover : alk. paper) | ISBN 9798855802078 (ebook)
Subjects: LCSH: Communication of technical information.
Classification: LCC T11 .T23 2025 | DDC 601/.4—dc23/eng/20250131
LC record available at https://lccn.loc.gov/2024042771

Miles Kimball would like to dedicate this volume to his parents,
Gerald Ward and Jean Kimball,
who have supported him across his long career.
Their kindness and generosity have made all things possible.

Hayley McCullough dedicates this collection
to their awesome parents—Stu and Cathy—
because they kept Hayley sane through this whole process.

Contents

Introduction

Miles A. Kimball, Hilary A. Sarat-St. Peter,
and Hayley McCullough

Establishing Technical Communication

Technical communication surrounds us. It's a chameleon, using script and symbol across media and changing shape to meet its purpose. At the most obvious and obviously public, we find endless instructional manuals or videos created by corporations and everyday users, teaching us how to do things both necessary and frivolous. But we also find private uses, such as when it directs surgeons which side to cut (*left* or *right*). For good or ill, technical communication organizes our existence. It tells us what button to push, what action to perform next, and where not to put our hands for fear of injury or worse. We carry technical documentation in the form of state and employer identification. Patients with medical implants rely on a "no MRI" tattoo to receive proper care if they are unconscious. (Of course, this benign example contrasts with the more malignant of technical communication tattoos practiced in concentration camps). We even wear technical communication, in tiny labels that tell us, in words and symbols, how to care for our clothes.

Technical Communication and Michel de Certeau

Technical communication happens at the point where institutional power and individual consent constantly struggle—the point where technical documentation speaks to users visually and where users respond conspicuously. Corporations use documentation to control users' actions and keep them in the black box of corporate control, while users—frustrated by the opacity of their experience with institutional technologies—ignore corporate control to make technology align with their desires and needs.

Technical communication often occurs as a by-product of industrialization, incorporation, or government. Rather than seeing technical communication as a product in and of itself, people tend to use the term in relation to a set of documents that explain how some other product could, should, or does work. As such, it functions as lexical (discursive) documentation, manifesting in three primary forms (alongside a myriad of subforms):

- Reports (what we did or what happened in the past)
- Instructions (what we should do right now)
- Proposals (what we want to do in the future)

These traditional documents make a set of rhetorical moves that relate more to the people who use technology; thus, people more frequently see technical communication as a complex set of practices that help accommodate people to technology or technology to people (Slack, Miller, & Doak, 1993). These practices, however, almost always present communication in terms of interactions between corporations, government agencies, and similar entities—in other words, the participation and prosecution of business.

Until recently, people have almost entirely framed this kind of communication in strategic terms, as described by Michel de Certeau (1984) in *The Practice of Everyday Life*. In that book, he examined what he considered the final unexplored corner of human experience: the everyday life of the common man. De Certeau analyzed elements and procedures that, while seemingly mundane, revealed fascinating aspects of human nature. And it is in this interest that we find the alliances between de Certeau's work and that of technical communication. And yet we have no evidence that de Certeau knew about technical communication before his death in 1986; however, we understand that technical communication was at its peak, leveraging an overlap between two significant trends in the field. At that time, technical communication was at a crossroads between its "brass age"—when technical communication was part of the large military buildup at the end of the Cold War aimed at outspending the Soviet Union—and its "glass age"—when technical communication shifted in response to the birth of the computer and video game industries (Kimball, 2017). Technical communication at this intersection flourished in a sort of intellectual boomtown. De Certeau (1984) might have called the noise at this intersection a "Brownian motion"; we might see it as the meaningful tones of a modem handshake. In some ways, de Certeau worked in the middle of this boomtown atmosphere, teaching in the 1960s and 1970s at UC-San Diego, during a time when the University California system received substantial government funding to build its computer expertise; the United States investment paid "Billions for Confusion" (Bishop, 1963)

Recognizing the difficulty in delineating the ensemble of procedures, such as operations and technical manipulations, de Certeau (1984) turned to two other French theorists, Michel Foucault and Jean Bourdieu. Foucault's interest in the self-regulating systems of power led him to consider how "tactics" within operations served as practices of consumption. Bourdieu's habitus examined how—in an economy of proper place and patrimony—learning interiorized structures and practice exteriorized achievements. For de Certeau, Foucault and Bourdieu articulated a "way of making" based on the ensemble of procedures involving place, strategies, and tactics. To understand how operations work, de Certeau suggested one should first "cut out" select practices and treat them as a separate population. This would allow for a redistribution of space for maneuvers, creating a play not entirely constrained by the original standards or structures. One should then invert the practices and knowledge, removing the control of their authorizers and temporarily allowing users to illuminate and use this knowledge to fit their individual needs.

For de Certeau (1984), this user-focused stage involved trickery: people could try to outmaneuver the steps or procedures directed at them, even though in doing so they might grow frustrated or lack the experience needed for effective action. In attempting to outmaneuver, users would combine things that did not "technically" belong together or that were not made to be changed or used in a manner beyond institutional design and control. *Bricolage*, as de Certeau (1984) called this practice, allows a transformation in which the consumers/users also become producers/makers in their act of making objects become something other than what the institution intended or designed them to be. Under bricolage is *la perruque* (French slang for "the wig")—another set of tactics that arises from the workplace and allow for popular techniques on pushing boundaries and reshaping rules (de Certeau, 1984). While bricolage allows for a transformation in user and object, *la perruque* appropriates time and material from work for personal use (more simply known as a cheat). Both bricolage and *la perruque* embody the ideals of tactical technical communication—where communication happens between peers rather than hierarchically at the whims of institutions.

Since de Certeau's treatise, users of technology have evolved into the "user-producers" (Johnson, 1998). They both consume strategic, institutional technical communication and create their own technical communication. These user-producers recognize the shift from institutional to extra-institutional settings and often place greater trust in the work of fellow amateur technical communicators than in that of professional tech writers hired by corporations. Institutional technical communicators excel at presenting how technology should function (at least from the corporation's viewpoint). In contrast, a fellow user is freer to explain how a product actually works; they face fewer barriers to demystifying a product. While a corporation typically resists alterations, a user-producer can share how they disassembled, molded, combined, adapted, rearranged, and fundamentally transformed the product according to their own needs. This kind of bricolage (though Johnson didn't term it as such) is central to the culture of sharing technological know-how through tactical technical communication.

The Tactical in Technical Communication

The emergence of *tactical* technical communication as a field of study has its origin in, as de Certeau would call it, a *perruque*. When Kimball (2006) published "Cars, Culture, and Tactical Technical Communication," technical communication as a field imagined its terrain as corporate, industrial, and governmental, and its objects, the manual and other instructional modes. This technical communication happens within organizations or between organizations and their members, constituents, or customers. Kimball posited that however important this institutional technical communication may be, it overlooks the technical communication performed by millions of people each day outside of and even counter to organizations.

This kind of technical communication has long existed but has grown tremendously with the opportunities afforded by the internet—particularly, people's ability to share technical information for their own purposes rather than on behalf of institutions. For example, several years ago Kimball learned how to turn wooden bowls on a lathe. It's a surprisingly complex task, requiring knowledge of wood (its species, hardness, cutting quality, moisture content, grain), tools (the hardness of the steel, the angle and shape of cutting edges), abrasives (including competing international standards for grit size), and physics (what happens when you push a piece of sharp steel into a log spinning at 1,500 rpm). In previous years, he might have relied on traditional sources of technical information such as the lathe manual, books, or courses at the arts center. However, his more recent guidance came from the myriad of YouTube videos created by more experienced and skilled woodturners. These instructional videos, while not always polished, represent remarkably effective pieces of technical communication. They're also typically free—offered out of pride of accomplishment and a willingness to share expertise and knowledge. And many users frequently turn to this kind of technical communication for guidance, instead of official, institutional documentation.

Effectively, everyone who has access to the internet has the potential to become a technical communicator, sharing their knowledge about technology with the entire world. Through platforms such as YouTube, Instructables, and web forums, even those with minimal resources can globally share information that was once the domain of professional technical writers. A common first step people now take when they encounter a technological problem is not to crack open the manual, search the help file, or even visit the company's online help, but simply to type the problem into Google and see what pops up. It's likely that someone else in the world has faced the same issue and found a solution they are happy to share openly. Manufacturers, recognizing the embedded expertise of their own user base, frequently harness or co-opt this expertise by creating open forums where users can share information with each other, rate the validity of the help they receive, and provide valuable metrics on the product and its use.

In 2006, Kimball wished to shift technical communication's analytical frame from subject-matter experts embedded in the industrial/governmental/corporate contexts to these user experts. And since the field didn't include space for this sort of consideration, Kimball created that space, justifying this new lens with de Certeau's theories on strategies and tactics, bricolage and *perruques*. He called this kind of technical communication "tactical" by building on de Certeau's (1984) distinction between strategies as the systems set up by institutions and tactics as individual actions that cut across or against institutional strategies (Kimball, 2006). Strategies refer to the actions that corporations, institutions, and agencies take to maintain control, while tactics refer to the actions individuals take to live their lives as they wish. Drawing de Certeau's (1984) ideas into the context established by Johnson (1998), Kimball's 2006 essay offered a transformative act of bricolage.

Kimball recognized that de Certeau's analysis of elements and procedures, while seemingly mundane, revealed fascinating aspects of human nature, and he used de Certeau's work to expand technical communication beyond its then-existing limits.

By *Tactical Technical Communication*, Kimball (2006) meant the technical communication that occurs outside and often against the grain of companies, organizations, government agencies, and other institutions—sites where we typically think of technical communication happening. If tactical technical communication seems too ambiguous, an alternative term many have used is *extra-institutional technical communication*. Tactical technical communication, while lacking a formal methodology due to the field's relative newness, employs unique technological narratives through which the user navigates the cracks of place or propriety of an institution and engages others to share their own version of the tactical story.

Since Kimball's essay, others have also drawn on De Certeau's concepts of "strategy" and "tactics" to articulate how technical communication works in everyday life. Through their work, we now see technical communication's nature and reach more fully. Ubiquitous, familiar, flexible, and usually useful, technical communication is a powerful social force. It is tactical, creating a layer of script on human experience and innovation. While de Certeau (1984) offered the field the terms to embark on that expansion, the field doesn't need de Certeau to continue it. Tactical technical communication has grown sufficient to justify its study.

The essays included in this volume draw on two decades of work on "tactical" or user-produced technical communication. Broadening the scope of prior research on "strategies" and "tactics," they examine how people can occupy many different roles—living and working as professionals, strategists, designers, tacticians, users, and everyday people. In particular, technical writers switch back and forth between these perspectives, often balancing the user's and the institution's interests to satisfy institutional protocols while also empowering people. Taken together, these chapters build a deeper, more nuanced understanding of technical communication, especially as it relates to individuals and institutions in society.

Part One: Reimagining Institutions

This section examines how people deploy tactical technical communication to influence, subvert and change institutional policies and procedures to the benefit of others. For instance, many professionals deploy workarounds and other tactics to better meet the needs of people that an institution is designed to serve. Responding to current tactical technical communication focusing on the communal or collective, Jessica McCaughey and Brian Fitzpatrick's chapter "In-Plain-Sight Tactics: One-to-One Advocacy in the Workplace" explores how employees at the Department of Veterans Affairs advocate for people within the institution's bureaucratic systems. Case workers use written documentation to "translate" veterans' stories into administrative language, a carefully crafted tactic to better serve vulnerable populations.

Similarly, in "LGBTQ+ Programming at the Local Library: Tactical Technical Communication as a Way of Addressing Gaps in Community Resources and Representation," Dannea Nelson and Emily January describe how librarians develop book displays and other programmatic initiatives to enhance LBGTQ+ visibility in local libraries. Such cases also illustrate how tactical technical communication can occur across various workplaces and disciplines, extending far beyond the scope of individual resistance to authority and playing a role in community and advocacy.

Unfortunately, some institutions adopt a welcoming posture while employing discriminatory policies that marginalize people. In "Exposing the "Actively Enforced" Policy: Tactical Technical Disruptake for Rhetorical Deinstitutionalization," Walker Smith examines how some religious organizations have changed their tone, but not policies, towards LGBTQ+ inclusion. This mismatch between what churches say and what churches do has sparked grassroots initiatives (or rhetorical deinstitutionalization as a means to produce technical guides) to document and publish evidence of discriminatory and controversial practices within religious organizations to provide more clarity around the potential risks present. Tactical initiatives, like tactical technical disruptake's ability to disrupt habitual operations and affordances, can help people avoid discrimination and provides the potential to provoke changes to church policies.

When tactics spread beyond the individual, they can make a collective impact on institutions and communities. In "Grassroots Activism and Tactical Communities: Examining the Poor People's Corporation in Mississippi in the 1960s and 1970s," Don Unger explores the use of tactical technical communication within a grassroots organization, the Mississippi Poor People's Collective. Archived technical documents illustrate how people used bricolage and other tactics to navigate encounters with white supremacist police officers, galvanizing broader efforts to promote social justice through resistance.

Also addressing issues in policing, in "The Streets Have Eyes: 'Cop-watching' as Tactical Engagement With Police Policy," Michael Knievel analyzes copwatching as a tactic originating from Black Power movements to promote police accountability. Drawing on observational and interview research with three cop-watching organizations, Knievel traces the ways in which copwatchers not only record events, but also seek to alter them, promote accountability, and improve citizens' experiences with police encounters through observations. Knieval analyzes the cop watchers' tactical technical communication through asserting citizen influence as an attempt to manipulate institutional for more equitable and just forms of policing practices.

Unfortunately, the use of tactics does not always empower people. In "Tactical Tech Comm and Failures in Crowdsourcing and Mass Production," John Sherrill highlights the limits of tactical action. Some "craftivist" responses to the Trump administration, such as the crafting of pussy hats, fell short of political and social objectives. Therefore, technical communicators must acknowledge the unique and contrasting logic to production of particular situations and through framing in order to fully take advantage of opportunities that may lead to devastating consequences around choice.

Other tactics succeed through alignment with institutional values and strategies. In "#WearAMaskNY: Public Service Announcements: Tactical Technical Communication During a Global Pandemic," Shannon Sarantakos and Sara Doan examine a state government's efforts to crowdsource the development of materials that promote mask wearing. Although the #WearAMaskNY was designed as a grassroots campaign, citizens who used professional production methods and equipment were more likely to have their message adopted in state-sponsored materials. This case illustrates how some tactics can depend on privileges such as education, wealth, and access to resources such as professional production equipment.

Corporations, too, must balance accountability to users with legal and financial objectives. In "Tactical User Research: How UX can (Re)Shape Organizations," Guiseppe Getto draws on a case study from the boating industry to illustrate how even the most "user-positive" organizations can prioritize the drive to make a profit over user empowerment and safety. They suggest ways that technical communicators and other user-oriented professionals might use technical communication tactics to convince organizations to empower users when competing interests are in play.

Furthermore, different stakeholders bring differing levels of power and authority. In "The Motivations of the Marginalized: Identifying and Navigating Hegemonic Factors in Petroleum Risk Communication" Joseph

Williams describes risk communication on an offshore petroleum rig, where multilingual and often marginalized crew members deploy tactics such as translation to communicate and amplify safety concerns. Williams examines the importance of communication in technical documents that need to consider cultural factors particularly when dealing with multicultural workforces. Such tactics not only promote safety and improve communication between management and workers, but they also allow workers from diverse backgrounds to create a sense of community.

Part Two: Transforming Society

The chapters included in this section examine the potential for tactics and tactical technical communication to not just change institutions and policies but to also transform society. First, transforming society requires reexamining history. In "Engaging Eurocentric Legacies in de Certeau's Thinking through a Self-Reflection on 'Queering Tactical Technical Communication,'" Avery Edenfeld and Steve Holmes unearth strains of Eurocentricism in de Certeau's theories, methodologies, and writing in order to consider how tactical actions for queer users are intersectional among others. By examining the Eurocentric roots of de Certeau's thinking, Edenfeld and Holmes offer a self-reflection of sorts, contemplating the tensions and connections of tactical technical communication to social justice and intersectional concerns. They highlight ways that technical communication scholars might work around these problematic elements, like the social limitations and lack of technical knowledge regarding BIPOC trans research in relation to tactical technical communication, to promote a more inclusive understanding of strategies, tactics, and other logics of production.

At the same time, tracing the evolution of tactics through tactical technical communication allows researchers to broaden the scope of historical inquiry. In "DIY Instructions in *BIKINI KILL*: A Feminist Historiographical Approach to Tactical Technical Communication," John L. Seabloom-Dunne examines tactical and technical aspects of a feminist punk zine that features DIY instructions on a variety of topics. When paired with a tactical framework, feminist theory and historiography enable scholars to further contextualize tactical practices by situating those practices within evolving structures of marginalization and resistance.

Similarly, in "*Our Bodies, Ourselves*: A Case for Metical Technical Communication," Kevin Van Winkle connects tactical technical communication to the ancient Greek concept of *metis*, or cunning intelligence. In particular, metis sheds light on the embodied nature of tactical action—an area of central importance to feminist technical communication scholars—but requires emphasis on the user-producer's corporeality. Van Winkle's chapter offers metic technical communication as a supplementary to tactical technical communication in its ability to subvert dominant culture and organizations whose strategies aim to control and manipulate all types of bodies.

Tracing the evolution of tactics through tactical technical communication can also help researchers understand how some movements turn violent. In "Armed Propaganda and the Ethics of Horrorism: Tactical Communiques from the Weather Underground," Brad Lucas draws on a case study of the complex tactical technical communication dynamics of the Weather Underground Organization, now categorized as a terrorist cell, to extend and expand de Certeau's focus on agency and resistance. Over time, the Weather Underground compromises its ethical bearings—eventually mirroring the institutional violence that the movement was constructed to resist. Lucas considers the ethical trajectories behind tactics that seemingly promote equality and justice as a form of resistance but through propaganda and extreme acts become horrorist agendas.

Often, tactics and tactical technical communication reflect the shared narratives, metaphors, and fantasy that define groups and movements. In "Hospitality at the End of the World: An Ideological Rhetorical Criticism of Tactical Technical Communication in *The Prepper Journal*," Ryan Cheek traces how preppers construct an apocalyptic vision of the future in which hospitality is earned by a select few who can demonstrate competency in survivalist practices. By exploring the intersections of tactical technical communication and ethics, Cheek's textual analysis of the apocalyptic rhetoric and ideology of prepperism that emphasizes enmity and tactical survival.

In many cases, understanding those narratives and metaphors opens possibilities for choice and reflection. In "Marx in the Digital Age: The Critical Role of Tactical Technical Communication in Contemporary Humanism," Sandy Brack develops and extends metaphors for twenty-first-century writing by presenting writing as a mirror that allows people not only to see and understand themselves but also to see their own reflections in consumer products. In this way, tactical technical communication allows people to move from "manufactured" consumer identities to a more authentic expression of the self-grounded, individual identity in tactical and creative action.

In "The Narrative Construction of Social Justice in Technical Communication Pedagogy," Tracy Bridgeford braids a personal narrative about teaching tactical technical communication through the lens of positionality, privilege, and power (Walton, Moore, & Jones, 2019) and with scholarship about the social-justice turn in technical communication. In a technical writing course, the study of fictional narratives about technology and society can generate possibility of imagined action—especially action toward creating more just, equitable, or sustainable societal structures.

Some tactical approaches, in contrast, emphasize individual autonomy over possibilities for collective action. In "Finding Agency Through Technical Communication: Privacy and Data Surveillance," Sarah Young and Jason Pridmore examine the representation of surveillance scenarios and tactics in popular YouTube tutorials about maintaining digital privacy. By focusing on individual, often limited actions that people can take to protect their own privacy, such videos address collective actions that people might take to resist the surveillance strategically of large governments and corporations. Young and Pridmore analyze how tactical spaces such as YouTube help users gain agency through resistance and challenge the problematic nature of data surveillance.

Acknowledgments

Miles Kimball would like to thank Coady Spaeth for her help on the introduction.

References

Bishop, M. G. (1963). *Billions for confusion: The technical writing industry*. McNally and Loftin.

de Certeau, M. (1984). *The practice of everyday life*. University of California Press.

Johnson, R. R. (1998). *User-centered technology*. State University of New York Press.

Kimball, M. A. (2006). Cars, culture, and tactical technical communication. *Technical Communication Quarterly, 15*(1), 67–86. https://doi.org/10.1207/s15427625tcq1501_6

Kimball, M. A. (2017). The golden age of technical communication. *Journal of Technical Writing and Communication*, *47*(3), 330–358.

Slack, J. D., Miller, D. J., & Doak, J. (1993). The technical communicator as author: Meaning, power, authority. *Journal of Business and Technical Communication, 7*(1), 12–36.

Walton, R. W., Moore, K. R., & Jones, N. N. (2019). *Technical communication after the social justice turn: Building coalitions for action* (First edition). Routledge.

Part 1

Reimagining Institutions

Chapter One

In-Plain-Sight Tactics

One-to-One Advocacy in the Workplace

JESSICA McCAUGHEY AND BRIAN FITZPATRICK

The study of tactical technical communication (TTC) thus far has focused on outsiders or those on the fringes of a system or institution creating and sharing workarounds, from patients grappling with complex diagnoses (Bellwoar, 2012; Holladay, 2017) to Volkswagen enthusiasts (Kimball, 2006). Yet in this investigation, we consider how those *within* formal institutions develop tactics and put them into practice. Michel de Certeau writes that "the space of the tactic is the space of the other" (1984, p. 37); yet in this chapter, the other is still an insider—one who in some ways acts as both navigator and translator, cunning (Dolmage, 2020) and advocating for vulnerable individuals by knowing the minutiae of a system well enough to technically stay within its confines, while, in many ways, working around it.

This chapter draws on interviews with nearly fifty professionals across fields whose work is not just unexpectedly technical but also, in many instances, employs unexpected, tactical workarounds. Participants in fields and roles as wide ranging as social work and nursing to marketing and even a lab manager working in an NIH-funded neuroscience lab recall ways in which they have learned or developed tactics to advocate for another person or group. This occurs when the professional realizes that traditional or expected institutional rules and strategies don't allow them to serve others in ways that they see as central to their work.

Here, we look closely at two professionals: a neurosurgery physician assistant (PA) working in an emergency room in a large, East Coast city, and a legal administrative specialist (LAS) at the Department of Veterans Affairs (VA) who fields calls from veterans, their dependents, and their surviving family members. Like most professionals, the PA and the LAS both rely largely on writing as part of their daily work. Specifically, these participants explicitly define their own writing largely as "documentation" and templated writing; however, through our conversations, we can see them making complex decisions and engaging in tactical technical communication on behalf of their clientele, despite the limitations of their writing modes and of the very systems to which they belong.

Each must work within "resource-constrained environments" (Rose, 2016) in order to help their vulnerable clients or patients navigate complex bureaucracies to receive care. In both cases, the technical documents these professionals produce are a foundational strategy of the institution: written records in an electronic system. Yet, the PA and the LAS turn these documents into genres that can be used tactically for advocacy in the name of patients or clients, some vulnerable in obvious ways (veterans with disabilities) and others who perhaps lack power or knowledge in the moment (emergency-room patients who would not know on their own how to secure a particular doctor's care). While they are neither the "enthusiasts" (Kimball, 2006, p. 74) nor "user-producers" (Edenfield et al., 2019; Holladay, 2017; Johnson, 1998; Kimball, 2006, 2017; Sarat-St. Peter, 2017) that we typically think of engaging in or driving TTC, these writers still take individual tactical action to subvert larger structures. They produce expected, "proper" documentation within the realm of their professional organizations, while using their savvy to make sometimes opaque systems work for their clientele in ways that still subvert expected uses.

Our exploration of the communication practices of these two professionals illuminates a kind of one-to-one, workaround advocacy through what we're calling "in-plain-sight tactics." In these close-quarters tactical scenarios, our advocates rely on intimate and personalized knowledge of their clientele as well as their expert knowledge of existing technicalities within complex systems to help the vulnerable successfully navigate these systems.

One of our goals is to provide a glimpse into TTC across fields as a first step to further research in unexpected or rarely studied institutions or organizations. "Technical" communication can be found just about everywhere—accounting and government policy, manufacturing and medicine, information technology and architecture—and within it, we can see tactical moves in the form of in-plain-sight tactical advocacy. It's our hope that such analysis will provide new, nuanced insights into technical workplace communication broadly across fields but also into the ways that documents and documentation systems are designed and employed and how such design sometimes fails the clients, patients, or other individuals they are meant to serve.

Beyond Communal Tactical Technical Communication

Much of the scholarship surrounding TTC to this point has illuminated spaces and exigencies of vulnerable or disadvantaged groups—particularly how inventive and collaborative communicators can—and should—be in problem solving when a system fails or excludes them. For instance, in medicine and related fields, we see explorations of TTC as it relates to specialized knowledge—and a lack of power—as it can be spread to a collective. Holladay (2017) offers a look at participants in online forums with particular psychiatric diagnoses attempting to interpret formal medical documents for and with one another. McCaughey (2020) similarly analyzes forum posts about "exclusive pumping," the underdocumented method for feeding infants when mothers cannot breastfeed. In these and other communities, we see extra-institutional communication channels form, wherein experienced lay people do the heavy lifting to help other lay people navigate complex tactics when an established system has failed or excluded them. In recent years, we've seen crucial insights about collectives and coalitions within systems, such as Walton and colleagues' *Technical Communication After the Social Justice Turn: Building Coalitions for Action*, which offers a set of "action items for redressing inequities" that includes recognizing and revealing systemic oppressions (Walton et al., 2019).

Below, we explore a form of tactical technical communication that fails to precisely align with traditional sightings of tactics; we see certain players within these larger systems who find themselves frustrated by their inability to change the bureaucracy—and so they begin making complex tactical moves under the radar, from the inside, sometimes against their training, and always of their own volition. Our PA and LAS are not working with user- or outsider-generated content or banding together with other like-minded individuals or those who share their challenges (Edenfield et al., 2019; Holladay, 2017; McCaughey, 2020). Neither are they creating or distributing information or advice counter to institutional strategies (Kimball,

2006; Sarat-St. Peter, 2017). Further, they break no institutional rules (Ding, 2009; Rose, 2016). Instead, these professionals advocate for their clients and patients tactically, in moments where "proper" strategic approaches might fail, while closely following institutional rules within the very systems the institutions have built in plain sight.

We consider these approaches in drawing on data from a larger, qualitative, IRB-approved study that focuses on ways that diverse professionals transfer writing knowledge and skills within their roles. Long-form interviews were conducted with nearly fifty participants, most in person, about the writing processes and products critical to their work and how they learned to be successful communicators in those roles. Representatives from many fields were interviewed, including workers in health care, government, business, arts, education, and the sciences.

We employed narrative theory as we developed the study, which, as Jones (2017) notes, provides "the researcher a unique sensitivity to participants' epistemological and ontological perspectives by tapping into their lived experiences. Ideally, participants guide the meaning-making process by sharing their own stories in their own voice" (p. 327). These narratives demonstrate what Van Ittersum (2014) refers to as an "account of craft" (p. 238) in his analysis of narrative in online "Instructables." Van Ittersum also speaks to "the story of the author's resourcefulness in overcoming an obstacle" (p. 242). This "resourcefulness" resides at the heart of the narratives of our two professionals.

Tactical Moves in Strategic Documentation

Our two subjects are resourceful in their employment of complex institutional expertise on behalf of vulnerable people who must interact with these larger systems. Such systems, with seemingly purposefully opaque rules and statutes, long wait times, notoriously unhelpful bureaucrats, are often inaccessible to the average user. This is at the core of the problems our PA and LAS address in their particular communication choices within their respective electronic systems. Each serve as both advocate and guide in the records they draft, despite such records being known simply as "documentation," rather than the persuasive texts they become in the hands of our expert tactical communicators as they work to secure the best care possible for vulnerable individuals who have neither the expertise nor the access to do so for themselves.

These tactical communicators operate in large-scale, resource-constrained institutions, both of which have reputations for being, at best, bureaucratic nightmares and, at worst, uncaring. Rose (2016) speaks to "resource-constrained environments" and defines resources as potentially including "physical as well as social infrastructure" and ranging "from robust bandwidth to electricity to literacy" (p. 433). The institutions that employ our professionals here are resource constrained in terms of funding, the reality of time constraints and medical expertise, or the necessary reliance on related institutions, such as the insurance industry and Congress. In this chapter, we examine these two cases in the context of institutional spaces dedicated to strategy and how our two interviewees subvert strategic expectations by using them to advocate for vulnerable patients and clients.

Case: Legal Administrative Specialist, Department of Veteran Affairs

While we may not immediately consider the VA health-care system as a limited-resource system in the same way we'd consider, say, water and food to be limited resources, due to years of underfunding, partisan attacks, distrust, and complaints about antiquated bureaucratic requirements (Stoddard et al., 2017; Tiefer, 2018; Turek, 2017; Bronstein et al., 2014), access to benefits for many veterans is uncertain. The LAS recounts his experience with clients who, without his tactical interventions, might never gain access to the benefits they deserve. His ability to help his clients get their medical benefits depends largely on his knowledge of what kind of application will and will not receive approval, as well as intimate and personalized details from his clients' histories. Sometimes, the line between receiving benefits and rejection comes down to extremely precise framing of a veteran's narrative:

> So there's something called a presumptive benefit, where if you were there and you have this disease, we assume it's because of this thing. An example of that would be exposure to Agent Orange. Another big example is exposure to contaminated water at Camp Lejeune. So if you have diabetes mellitus, and you were in Vietnam during the time when they were spraying Agent Orange, then we can assume that there is a causal link between the two. So there are certain special subsets where that applies. However, for that to work, you have to be able to show that you were in Vietnam during that time, or another

> affected area. So with that being said, if they don't know how to link—like if they don't know how to show that they were in Vietnam during that time, which will be difficult for certain groups, say naval—if you're far enough out at sea, we might not be able to consider you presumptive. However, if you were inland when you were dropping troops off really close to the shore, we may still be able to do that. You have to be able to show things like where your ship is stationed and things like that. There was one where he'd mentioned a newspaper article that mentions people in his unit. I helped him kind of pull that together, show where somebody in that unit was–that there was an accident or something that occurred, so that I could help place him where he was.

Our LAS knows that simply employing the strategies of the VA system—documenting a veteran's personal "bite-sized information," a concise and accurate representation of their narratives—would be an acceptable performance in the eyes of his superiors. He is judged in monthly reviews not by how many clients he helped obtain benefits, or the level of detail with which he presents their case, but by the precision with which he has followed protocol: "Did you write it correctly? And did you send it to the right people?" In the instance that he merely took accurate notes from, in his example, a sailor stationed off the coast of Vietnam during the Vietnam War, he would likely see the veteran denied presumptive benefits. His institutional expertise and *metis*, though, allow him to anticipate a rejection and translate the veteran's story; he recognizes vital specifics and asks probing questions to uncover pertinent details, such as if "[they] were inland when [they] were dropping troops off or by finding a newspaper article that could serve as record that they were there." He notes that "[veterans] might not even realize the important thing that they just kind of said in conversation, how important that is, and I might need to translate that." The LAS serves as both navigator and translator of this complex system on his clients' behalf, employing tactical approaches to check the right boxes and articulate the right details in order for his clients to receive proper benefits, even when his client might be failed by the absence or imprecision of traditional (strategic) records.

Case: Physician Assistant, Neurosurgery, Large Urban Hospital

Similarly, in our interview with a PA in neurosurgery, we see her use her expertise to guide vulnerable parties ("trauma patients [. . .] that have had head injuries, spine injuries, or spinal cord compression [as well as] patients in the ICU, either before or after surgery") through the complexities of the hospital system. She does not see herself as a technical communicator, describing her writing primarily as "documentation and billing for the hospital, and just as what is legally required in health care," and adding that "most of our writing is actually in template form." Yet she actually performs quite a bit of persuasive advocacy on her patients' behalf to help get them, in her view, the best medical care possible. She states:

> I think the way that I approach it is, "How do I shape [the patient's] story into something that's going to catch someone's attention?" [I]f I get a call for a consult to say, "Hey this patient has some kind of an issue and it looks like they have a fracture in a bone near the ear," I'm immediately checked out thinking, "Why are you calling a neurosurgeon for this? We don't take care of this. I'm not interested." So, same thing if I'm trying to talk to a medicine doctor. I'm trying to frame my note that would be appealing to them to say, "Hey this is exactly why we need you, and this is why we hope that you're going to accept our patient "because [. . .] once a surgical problem is managed and taken care of, you want to transfer your patient to a doctor that can better take care of their medical needs. [. . .] So you want to try to kind of frame the patient as, "Oh this is a really interesting medical patient now that we're done with the surgical part of things." So having to write something in a way that's going to make it relevant to other people and catch their attention is a big challenge in writing and I think it's a challenge that's kind of fun to try to do.

And when asked about why specialists might reject taking on a patient postsurgery, she says: "[They get] bored too, but also protective of their workload because, you know, I'm maybe seeing thirty patients on my service and [. . .] I'm having five other people try to give me other patients, so I'm thinking, 'Okay, do I really want this patient that's not at all really relevant to me?' " Because of her expertise and experience, with both the institution and the particular doctors within it, she knows that without a good "story" and proper, compelling framing, patients risk not getting access to the best or most appropriate doctor for their aftercare. Here, the formal institutionally accepted practices are easily achieved—basic historical documentation of the patient's experiences and writing "what is legally required" to cover herself and her institution—yet with that low bar met, the patient would be left to the whims of the hospital system regarding under whose medical care they'll proceed. To get a patient in her care accepted by a preferred specialist, someone she believes will offer the best and most appropriate care for that particular patient in that particular context, she must channel her expertise—including her knowledge of the patient, her knowledge of the specialist, and their workload and interest—to employ a strategic approach (de Certeau, 1984) and "shape the patient's story" in a way that the patient could never possibly advocate for themselves. Her institutional knowledge, combined with her particular understanding of her patients' needs, allows her to see the opportunity for tactical approaches to care.

Driven by Empathy: In-Plain-Sight Advocacy

In other explorations of TTC, often a person in need who has been failed by an institution seeks out tactics as a product of necessity. Sarat-St. Peter (2017) states that tactics allow communicators to trespass or "poach," in a way, on formal institutions or systems; they allow "people with the flexibility to adapt practices and procedures to the affordances and demands of a given moment" (p. 80). In this traditional view, the user moves from vulnerable person to tactician through these trespasses, empowered either through newfound access to knowledge or even simply through a solution or workaround to the problem.

However, our tacticians, the PA and LAS, are not the vulnerable parties here at all; they are, to varying degrees, in positions of power. Their clients and patients exist outside and at the mercy of their systems, which are so vast and perplexing that it would be easy for them to fall through the cracks and receive a poorer quality of care. Our tacticians are "trespassing" intra- and extra-institutionally, on the strength of their knowledge of their respective system ecologies, acknowledging and anticipating the cracks, not only to recognize the need for, but also to provide flexibility and tactics on behalf of, their vulnerable clientele.

Further, the patients here do not transform from vulnerable person to tactician as this work is done on their behalf; in fact, they might not even be aware of the pitfalls of these complex systems nor the possibility of failure or consequences thereafter. The veteran client may have no inkling that without a certain newspaper clipping or a particular line about their ship proximity to Vietnam being presented in their narrative, they risk being denied medical benefits. Similarly, the neurosurgery patient likely doesn't know that one doctor is better suited to their health concern than another. These seemingly small tactical approaches are not widely applicable. Rather, they are individualized to particular personal histories and needs. Advocating for them requires not only expert knowledge of the margins of the system but also very particular knowledge about the person. This isn't the type of tactical advocacy that can be performed anonymously on a message board. Rather than working within communities or collectives, these are individuals trying to help other individuals "up close," whether in person or remotely. A pure outsider could not gain the necessary knowledge, as it is, first, so specialized and, second, so very particular—customized, even—to the individual client or patient. Therefore, the outcomes of these tactical moves in the documentation are not communally applicable. Unlike anonymous strangers on an internet forum sharing tactics, these professionals will not likely distribute insight into their tactics. In fact, they may be the only person to ever employ them (and perhaps the only person to know that they have been employed), albeit likely in some form for a variety of patients and clients over time. The tactician is not necessarily empowering anyone with knowledge, nor are they transferring access in the way we see elsewhere (Kimball, 2006; Colton et al., 2017; Edenfield et al., 2019; McCaughey, 2020; Holladay 2017). The PA and LAS simply "assign" the vulnerable person a tactical approach that is possible within the institution's existing system to get them the care that they need, throwing a coat over a puddle their client or patient doesn't see. Through the advocacy work of our expert tacticians, the patients move from vulnerable to cared for.

These customized tactics depend not only on close, personal knowledge of their clientele but also on precise awareness of the complexities of these systems. The tactics are neither "exploits" nor loopholes; in fact, our tacticians are using the systems exactly as they are meant to be used. However, to the average user, these details or "rules" would seem to be concealed. While it appears that these two interviewees are employing strategies, they are, in fact, employing what we call "in-plain-sight tactical communication."

It's important to note that because these employees are extending their knowledge and power by utilizing these moves "properly," and in plain sight, they are not putting themselves at risk in doing so. They are working in the cracks, as de Certeau (1984) calls them, but not explicitly breaking the rules. The PA, for example, explains that her writing is always in template form within the medical records and is strictly documentation or "just . . . what is legally required in healthcare"; yet she writes persuasively, however subtly, on behalf of a patient to get them the best possible care after they leave her operating room. The Veterans Administration employee's documentation doesn't break formal rules either but does require more effort—more time, more digging through records and, at times, even old news clips, to find ways to appropriately "document" in favor of obtaining needed benefits for veterans. Neither communicator breaks explicit rules, but they spend more time and effort than necessary for their own workplace success to help their patients and clients. At a glance, it may not seem like taking the time to consider and write a few extra lines on a patient record or asking a few more probing questions to a veteran would be a significant increase in time or effort, until we also consider the size of the systems in which these tacticians operate and the volume of user-clients with whom they interact and for whom they advocate.

We argue that the driving force for them—and so many others working around barriers within institutions—is empathy. In online forums, anonymous users sharing their exclusive pumping experiences use their narratives not only to inform and help other women yield more milk when traditional avenues of care fail them but also to demonstrate and share empathy and camaraderie with fellow forum users (McCaughey, 2020). For the PA and the VA LAS, empathy appears, again, to propel their tactical actions. When their written documentation is evaluated, its success or failure has nothing to do with the "help" the employees' writing provided. The LAS is not praised for getting veterans more or fewer benefits; rather, he is judged on the accuracy of his documentation of the conversation. He relays that his monthly reviews are about following a set of steps, and "when you wrote it did you write it correctly? And did you send it to the right people?" The PA has already completed the seemingly most essential part of her job—caring for the patient and performing surgery—when she writes up the documentation that will accompany the patient to the next stage of his or her hospital stay. Her success is judged by, in her words, whether the record will "hold up in court." The PA's own success has nothing to do with whichever doctor the patient ends up seeing next, and yet she makes these efforts to direct their path to the care provider she thinks will serve them most effectively, presumably out of empathy and care. If the bar (strategy) is accuracy and precision, they're passing it regardless of whether they embark on these intimate tactical actions. Because they have so much knowledge about their respective institutions and the systems within them, these communicators can foresee the potentially negative outcomes invisible to their patients and clients. To cut off that outcome, they devise tactics and adapt procedures to help those who need their insight—whether they know it or not—to ultimately allow these people to receive the benefits that would be otherwise out of grasp. It's more work for the tacticians—work that largely goes unrecognized.

Studying In-Plain-Sight Tactics to Better Serve Vulnerable Populations

In their book, *Technical Communication After the Social Justice Turn: Building Coalitions for Action*, Walton and colleagues (2019) offer a set of "action items for redressing inequities" that includes recognizing and revealing systemic oppressions (p. 134). They offer the collective as a viable method toward equity. But what happens when a large, bureaucratic system's flaws that affect vulnerable people are already well known, such as in the United States health-care system or the systems through which veterans should be able to gain support and services? One might argue that a PA or LAS—even in coalition—would make little headway in the larger conversation about these oppressive systems. And yet, the two communicators we focus on here find ways to use their complex knowledge about these systems, working within them while advocating for vulnerable patients and clients. In an ideal world, the ability to advocate for those in need of support would be built into these systems. Or, at least, there would be mechanisms for effectively reporting those impediments—a less than stellar doctor, uneven military documentation in wartime—to allow for advocacy that is obvious and even required. Walton and colleagues (2019) clearly state that organizational policy should be a site where TPC scholars consider communication and procedural justice. They write, "When procedures are just, they pave the way for further conditions of justice by offering fair mechanisms by which to evaluate conditions, determine value, and set goals" (p. 41). And yet in both of the larger systems explored here, such inequities and barriers are known, and, for many reasons, go ignored and unresolved.

Further, as Walton and colleagues (2019) write, "historically, lived experiences and bodies have not been understood as central or important sites of study for TPC" (p. 125). We are, of course, witnessing a shift with the social justice turn specifically. This turn has spoken to systems of oppression, largely around race, and while we want to be particularly clear that we do not consider the tactics and actions of our participants to fall under "social justice," we do see clear advocacy, on an individual level, for vulnerable individuals. Here, too, our participants are decidedly focused on "bodies" and "lived experiences," through the lens of injury or disease and military history, respectively. For the PA, this includes neurosurgery patients with a limited knowledge of the larger American health-care bureaucracy and the specialties, skills, and experiences of doctors in a particular hospital. For the LAS, this includes American veterans with physical or mental ramifications from their time in the Armed Services, seeking benefits from a large, resource-constrained government agency.

While the two study participants we focus on in this chapter are not what we might traditionally think of as technical communicators—despite the nature of their communication being quite technical—they certainly perform tactical technical writing. De Certeau (1984) writes that tactics and those who seek to employ them "must vigilantly make use of the cracks" (p. 37). For these two participants, the cracks are language and details. Within their formal records—medical and governmental—we see these individuals advocating for others through subtle persuasion and by inquiring beyond what is necessary to find details needed to gain support and services. Further, we are not arguing that these two technical communicators are exceptions or exceptional, despite their displays of empathy and advocacy in resource-constrained environments; rather, they illustrate what we see as common—if often unidentified—moves that so many communicators make as they work around diverse institutional systems to support and advocate for clients, patients, and others who lack the background, knowledge, or access to advocate for themselves.

Gouge (2017) writes that "though it might make some nervous to think of clinical encounters in health and medicine and their documents as improvised," in fact it is in acknowledging such improvisation that we allow "the existing complexity to be made visible" (p. 429). Explorations like this one explore not only the in-plain-sight advocacy and tactical work these professionals perform, but they also expose the poorly designed systems that fail so many—not only because of a lack of resources but also due to the ways in which "documents" are thought of and designed. It is within these constraints that Rose (2016) tells us we might find opportunities "to examine tensions that occur between human intentions, actions, and needs and the systems, resources, and technologies that mediate our worlds" (p. 433). We agree, and we too see these spaces and the writing in them—medical or government-agency electronic documents, but surely many, many others too—as crucial to potentially coming "to see design needs in a different light" (Rose, 2016, p. 443). It seems clear that these systems can inhibit the people they are intended to serve. It's in examining the tactics of communicators like those included here that we might begin to better understand how such advocacy occurs and how institutions might better align their missions, systems, documents, *and* the expectations of professional communication across all three.

We might look further into these complex bureaucratic systems, seemingly designed to keep out their own vulnerable users, by reexamining not only the ways in which these ecologies might become more transparent and accessible to users, but also how these same systems might better assist and incentivize those who work within them to guide their clients. We believe these kinds of tactical moves occur in communication across industries and hierarchies. While we selected these two interviewees specifically because of the vulnerability of those they serve, they are not the only instance of this kind of in-plain-sight tactical communication that we have encountered in our work. For example, a certified public accountant notes in one of our interviews that she "tell[s] the saddest story the truth will allow" in a letter to the IRS on behalf of a client, to protect them from facing stiff tax penalties. These kinds of tactics are central to user navigation of complex systems, even if the tactics are unspoken and go largely unnoticed by both user and system alike. In performing analyses like this on different types of documentations and writing, we might discover points of inaccessibility and struggle not just for professionals working in resource constrained environments and writing in character-limited electronic documentation systems but also for the patients, clients, users, and potentially vulnerable clientele, who rely so much on the products and services of these institutions, but are, unfortunately, so often failed by them.

References

Bellwoar, H. (2012). Everyday matters: Reception and use as productive design of health-related texts. *Technical Communication Quarterly, 21*(4), 325–345. https://doi.org/10.1080/10572252.2012.702533

Bronstein, S., Griffin, D., & Black, N. (2014, June 24). *Whistle-blower: VA deaths were covered up*. CNN. https://www.cnn.com/2014/06/23/us/phoenix-va-deaths-new allegations/

Colton, J. S., Holmes, S., & Walwema, J. (2017). From noobguides to #OpKKK: Ethics of anonymous' tactical technical communication. *Technical Communication Quarterly, 26*(1), 59–75. https://doi.org/10.1080/10572252.2016.1257743

de Certeau, M. (1984). *The practice of everyday life*. University of California Press.

Ding, H. (2009). Rhetorics of alternative media in an emerging epidemic: SARS, censorship, and extra-institutional risk communication. *Technical Communication Quarterly, 18*(4), 327–350. https://doi.org/10.1080/10572250903149548

Dolmage, J. T. (2020). What is metis? *Disability Studies Quarterly, 40*(1), Online Publication. https://doi.org/10.18061/dsq.v40i1.7224

Edenfield, A. C., Holmes, S., & Colton, J. S. (2019). Queering tactical technical communication: DIY HRT. *Technical Communication Quarterly, 28*(3), 177–191. https://doi.org/10.1080/10572252.2019.1607906

Gouge, C. C. (2017). Improving patient discharge communication. *Journal of Technical Writing and Communication, 47*(4), 419–439. https://doi-org.mutex.gmu.edu/10.1177/0047281616646749

Holladay, D. (2017). Classified conversations: Psychiatry and tactical technical communication in online spaces. *Technical Communication Quarterly, 26*(1), 8–24. https://doi.org/10.1080/10572252.2016.1257744

Johnson, Robert R. (1998). *User-centered technology: A rhetorical theory for computers and other mundane artifacts.* SUNY Press.

Jones, N. N. (2017). Rhetorical narratives of black entrepreneurs: The business of race, agency, and cultural empowerment. *Journal of Business and Technical Communication, 31*(3), 319–349. https://doi.org/10.1177/1050651917695540

Kimball, M. A. (2006). Cars, culture, and tactical technical communication. *Technical Communication Quarterly, 15*(1), 67–86. https://doi.org/10.1207/s15427625tcq1501_6

Kimball, M. A. (2017). Tactical technical communication. *Technical Communication Quarterly, 26*(1), 1–7. https://doi.org/10.1080/10572252.2017.1259428

McCaughey, J. (2020). The rhetoric of online exclusive pumping communities: Tactical technical communication as eschewing judgment. *Technical Communication Quarterly, 30*(1), 1–14. https://doi.org/10.1080/10572252.2020.1823485

Rose, E. J. (2016). Design as advocacy: Using a human-centered approach to investigate the needs of vulnerable populations. *Journal of Technical Writing and Communication, 46*(4), 427–445. https://doi.org/10.1177/0047281616653494

Sarat-St. Peter, H. A. (2017). "Make a bomb in the kitchen of your mom": Jihadist tactical technical communication and the everyday practice of cooking. *Technical Communication Quarterly, 26*(1), 76–91. https://doi.org/10.1080/10572252.2016.1275862

Stoddard, C., Bonella, B., & Segovia-Tadehara, C. (2017). Veterans and the Department of Veterans Affairs. *Journal of the Utah Academy of Sciences, Arts & Letters, 94*, 359–372.

Tiefer, C. (2018, June 8). Veterans sustain two serious defeats from Trump and the House to VA health care. *Forbes*. https://www.forbes.com/sites/charlestiefer/2018/06/08/veterans-sustain-two-serious-defeats-from-trump-and-the-house-to-va-health-care/

Turek, M. E. (2017). "To care for him who shall have borne the battle:" Expanding the choice card program to provide for those who serve. *Kansas Journal of Law & Public Policy, 26*(2), 215–235.

Van Ittersum, D. (2014). Craft and narrative in DIY instructions. *Technical Communication Quarterly, 23*(3), 227–246. https://doi.org/10.1080/10572252.2013.798466

Walton, R. W., Moore, K. R., & Jones, N. N. (2019). *Technical communication after the social justice turn: Building coalitions for action* (First edition). Routledge.

Chapter Two

LGBTQ+ Programming at the Local Library

Tactical Technical Communication as a Way of Addressing Gaps in Community Resources and Representation

Dannea Nelson and Emily January

The work of tactical technical communication (TTC) happens across disciplines and workplaces, as those who engage in technical and professional communication (TPC) use rhetorical knowledge to accomplish various goals. Examples include resisting gender narratives at work (Petersen, 2018), interpreting psychiatry documents online (Holladay, 2017), and learning hormone self-administration for gender transitioning (Edenfield et al., 2019). All such examples point to TTC's role in social justice; such work is political and has implications for human rights and dignity.

One example of TTC's political nature is the case study in this article. During author Dannea Nelson's undergraduate program, she used TPC to tactically address a lack of LGBTQ+ programming at her workplace, the library. Her research was unsolicited, outside of paid duties, and focused on a topic of concern to her. She knew, in her conservative state, "that LGB adolescents' relationships with their parents are often challenged, particularly around the time of disclosure of sexual identity or 'coming out'" (Ryan et al., 2010, p. 206). Further, "from 2016 to 2018 . . . [State] had the fifth-highest age-adjusted suicide rate in the U.S. . . . In 2018, suicide was the leading cause of death for [state residents] ages 10 to 17" (IBIS, 2019).

This article presents Nelson's TTC surrounding the above problem and information about LGBTQ+ library programming from a TPC perspective. As an openly gay woman, Nelson's TTC allowed her to gain a deeper understanding of the narratives she collected—"an 'insider-outsider' perspective" (Mehra & Braquet, 2007, p. 545)—and to develop a mindfulness of research ethics and community needs. Unfortunately, her results received little interest from the library's administration. While disappointing, this was not surprising, as she found a lack of concern for and resistance to the LGBTQ+ community from management during her research. Such attitudes pervade our current political discourse. For example, Florida's "Don't Say Gay" law and similar legislation in twenty-three other states, including book bans, are attempting to silence LGBTQ+ voices and topics. Therefore, this chapter highlights the importance of TTC within community organizations and workplaces, the need for visibility, and the possibility of failure. When the community organization or workplace is uninterested in what TTC uncovers, where does it go? This article explores advocating for LGBTQ+ resources, the techniques and ethics involved in TTC, and questions that arise from disseminating TTC.

Literature Review

As it stems from de Certeau's strategies and tactics (1984), TTC has been well outlined in the literature (Kimball, 2006; Ding, 2009; Kimball, 2017; Sarat-St. Peter, 2017; Petersen, 2018). Of interest in this particular study is thinking of TTC as "something that makes lives better" (Kimball, 2017, p. 6). Further, TPC research ought to "bring change . . . [and] network and build voice for those traditionally silenced" (Mehra & Braquet, 2007, p. 543). LGBTQ+ lives are in need of attention in TPC research and in library programming. The American Library Association's (ALA) Bill of Rights states that "library resources should be provided for the interest, information, and enlightenment of all people of the community" (Stringer-Stanback, 2011, p. 1). Due to activist librarians during the 1970s and 80s, led by Barbara Gittings and the ALA Task Force on Gay Liberation, changes occurred in the cataloging of LGBTQ+ texts within the Library of Congress and libraries across the globe (Adler, 2015, p. 479). It was through TTC "that activist librarians were able to challenge a large cultural government institution and succeeded in changing the terms by which we come to knowledge" (Adler, 2015, p. 483). Gittings and others compiled bibliographies, created the Stonewall Book Award, organized speaking events, devised a new comprehensive subject scheme, and corresponded with the Library of Congress. These actions included documentation to make a difference in the way libraries operated. Faber (2002) noted that "change, as an identity function, occurs as a communicative project. This means that change occurs through the manipulation of specialized forms of language" (p. 29).

Even with such events, libraries continue underserving members of the LGBTQ+ community, especially youth (Mehra & Braquet, 2011; Montague, 2015; Robinson, 2016). At-risk youth turn to libraries for information about their identities as "they often lack community role models and honest perspectives on what it means to be homosexual" (Hughes-Hassell et al., 2013, p. 12). Further, they may be "forced to patronize public libraries because of a total lack of support from other social groups and services" (Robinson, 2016, p. 163). Such knowledge seeking is not uncommon among youth, as girls in the twentieth century often sought pamphlets and books for information about menstruation as they felt unable to approach their parents for such information (Brumberg, 1998).

Moreover, the first experiences with rejection for LGBTQ+ adolescents are often in their own families (Robinson, 2016, p. 163). There is a higher likelihood that LGBTQ+ adolescents experience bullying and violence perpetrated by their peers (Sang et al., 2020, p. 2). While nationwide opinion polls and social scientific studies indicate a "decidedly more 'liberalized' attitudinal trend toward gay men and lesbian women," these positive trends do not explain the "discrimination experienced by sexual minority children often occur[ing] regularly" (McCutcheon & Morrison, 2021, p. 113). Such experiences result in LGBTQ+ youth being at high risk for depression, anxiety, homelessness, substance abuse, and suicide (Hughes-Hassell et al., 2013, p. 2). These outcomes are not surprising considering the perpetual stress of trying to cope with being young and LGBTQ+ in "heteronormative and cisnormative environments" (Sang et al., 2020, p. 2). LGBTQ+ youth must have visible examples and develop successful rhetorical strategies for dealing with the conflicts that education, community, and society impose upon them (Peters & Swanson, 2004).

While youth in urban areas may have numerous resources to look to, adolescents in rural areas may not. Generally speaking, they almost always have access to a library. Gittings "believed that working with the Task Force on Gay Liberation to improve access to gay-positive fiction and nonfiction in the library was essential" in getting materials into the hands of those who needed them (Adler, 2015, pp. 482–83). Providing access to such books and other materials and texts is well and good; however, libraries should be more than a repository. For LGBTQ+ youth, libraries can also serve as a respite from bullying and violence and a place to create a community.

Despite the need for libraries in LGBTQ+ lives, some librarians have ignored and excluded LGBTQ+ references, maintained the status quo, used "delaying and strategic actions of diverting attention or bureaucratic procedures," and made "token gestures that do not make real changes" (Mehra & Braquet, 2011, p. 409). They may also "sidestep . . . responsibility, suggesting that LGBTQ+ and gender variant youth do not live in their communities" (Robinson, 2016, p. 164). This erasure makes LGBTQ+ patrons and employees a "seemingly absent user group" (p. 164). Such work on the part of librarians is strategic, what Feenberg (2002) called "operational autonomy" or "the power to make strategic choices among alternative rationalizations without regard for externalities, customary practice, workers' preferences, or the impact of decisions on their households" (pp. 75–76). The librarians use strategies that maintain institutional power; LGBTQ+ programming requires not only public outreach but also institutional change. Such change can happen through "reactive autonomy" or "margin of maneuver." This is "[a]ction on the margin [that] may be reincorporated into strategies, sometimes in ways that restructure domination at a higher level, sometimes in ways that weaken its control" (Feenberg, pp. 84–85). Such maneuvering can happen in the TPC genres of the library that shape how users interact with the institution. These genres include policies and procedures, board guidelines, fundraising initiatives, displays, brochures and other public-facing printed materials, and style guides for printed materials and online pages. There is always room to maneuver or tactically act, especially from a TPC standpoint, when we come up against power structures.

Due to erasure, LGBTQ+ interests should be represented in the genres of library TPC. Cox's (2019) data revealed that "queer rhetorical practices in workplaces . . . not only can help professional communication studies to more fully understand how meaning is made across professional spaces and situations but also can help LGBT professionals to be better understood and more successful in their professional aspirations" (p. 3). Through TTC, it is possible to upset erasure strategies within libraries by empowering employees, allowing them to see themselves as agents of change. Visible, public-facing TPC genres, such as displays and brochures, and invisible genres that train library employees on their duties to the public reinforce the identity of the library as safe. "Resources can provide self-affirmation, offer characters with which to identify, and decrease the feeling of alienation" (Hughes-Hassell et al., 2013, p. 4). TTC has a role in challenging institutional power. The need for workers to engage in TTC has been established, but what happens when a library employee engages in such work?

Methods

This chapter uses autoethnography to reflect on a research project that Nelson, a graduate student, conducted as an undergraduate in her workplace. Autoethnography "seeks to describe and systematically analyze . . . personal experience . . . in order to understand cultural experience" (Ellis et al., 2011, p. 273). We looked over the products and processes of Nelson's research, noting her experiences in TTC. We crafted her research and what she remembered as a narrative. We have anonymized all data to protect the identities of Nelson's coworkers.

Nelson conducted interviews and distributed a questionnaire in order to gather data. She did so as part of a final project in an undergraduate research-methods class taught by coauthor Emily January, and her research was covered under the IRB exemption for research conducted during coursework. When her research was conducted, she did not intend to publish her work, and it did not fall under the jurisdiction of the IRB. That said, our method is to analyze her process of writing and researching tactically within her workplace, not to analyze the data she collected. Our research method is reflexive/narrative ethnography, in which "the ethnographer's backstage research endeavors become the focus of investigation" (Ellis et al., 2011, p. 278). In that sense, the case study serves as data itself, data that illuminates the process of TTC and how it can be used for addressing social and cultural problems.

Case Study

In early 2019, Nelson carried out a research project for an undergraduate TPC course using the county library as her research site. Her interest was in programming aimed at LGBTQ+ youth and how such services might help to combat bullying and prevent suicide. As part of her employment, Nelson was asked to help frame the justification of an LGBTQ+ book discussion for teens called Lavender Literacy. Similar programs had been suggested in the past and had been denied, so there was a feeling of urgency to get approval. Over the past several years, the library had sponsored few programs for adults.

After hearing anecdotal evidence about the rejection of previous programming suggestions, Nelson started to wonder: Why is the library lacking LGBTQ+ programming to educate the public and provide resources for the LGBTQ+ community, specifically youth? As she spent time on the ALA's website and collected stories from colleagues, Nelson determined to find a resolution acceptable for the conservative community and the library system to offer programs that benefit the most vulnerable community members—teens—regardless of their self-identity. Nelson found herself uniquely positioned as a participant researcher with training in TPC, as a library employee and as an openly gay woman.

She conducted observational interviews with library staff in March 2019. Since the staff were already working on the Lavender Literacy program, the interviews were an organic part of discussions. Based on staff responses, Nelson found that there was always resistance from administrators when LGBTQ+ programs were suggested. Staff members who were interviewed agreed that the library should offer more programming surrounding education and support for LGBTQ+ patrons.

After hearing these opinions, Nelson created and distributed a questionnaire in April 2019 for county library employees with the library director's approval. The questions follow:

- Which LGBTQ+ and gender variant programs have been offered?
- What has the public's response been to those programs?
- What other resources are available for LGBTQ+ and gender-variant youth in the county?
- How do you, as a library employee, self-identify?
- How do you feel about the programming currently offered?

The questionnaire revealed several interesting pieces of the LGBTQ+ programming puzzle. Nearly one-third of questionnaire respondents selected at least one identity that was nonheteronormative or gender variant. Nearly half of responding employees believed library programming to be diverse.

Yet the history of the library's LGBTQ+ events tells another story. The first LGBTQ+-themed event sponsored by the library occurred in 2009, a screening of *Edie & Thea: A Very Long Engagement* to celebrate LGBTQ+ History Month. The showing led to angry phone calls directed at the organizing staff member. The same staffer hosted another film screening in October 2012, the PBS-produced *American Experience: Stonewall Uprising*. The library announced the event in a newspaper, and it did not generate an outcry. The next event was Rainbow Reads, an adult program organized by the same staff member, publicized and held every October since 2014. No member of the community has ever attended the event; however, the staff member has continued to put it on the calendar "because it's important to keep offering it" (personal interview).

Yet, respondents acknowledged that negative attitudes about LGBTQ+ programming prevented further programming, especially for youth. One respondent said, "Our public library system serves a very conservative patron base." One employee was told that "there's no point in planning it [LGBTQ+ programming] because people won't come." Yet that did not seem to be the only reason for a lack of programming. One respondent articulated frustration, stating: "I get frustrated at push-back related to and restrictions placed on programs that are intended to show the library as diversity-friendly, particularly as related to DGS [Diverse Genders/Sexualities]/LGBTQ+ communities because it may cause conflict or be seen as controversial. The majority in a community don't always need to be catered to at the expense of minority communities. The emotional/moral comfort of the majority is not more important than the health, safety, well-being, and inclusion of nonviolent minority groups." Employees are aware of the administration's unwillingness to offer programming. While the rhetoric is that there is a lack of interest from the community, interviews exposed the administration's resistance. They want to avoid controversy, without considering their ethical responsibilities to patrons, especially those of vulnerable populations.

Nevertheless, the lack of participation in LGBTQ+ programming is a concern, and the administration uses it as an excuse. One respondent shared, "I think the library does a great job of planning diverse programs. The problem we run into is that people do not show up for programs with more sensitive topics. We have held many programs that focus on race/racism during nationally recognized events like Black History Month, Hispanic Heritage Month, etc." This raises concerns about library resources.

Should the library continue to plan and execute events when there is no visible audience for it?

Several suggestions for LGBTQ+ programming came out of Nelson's research. One respondent suggested that the library partner with another county library that has a market for such programs, sharing expenses, labor, and patrons. It might address the lack of attendance; those in their county who are interested in LGBTQ+ programming might be more likely to attend an event partnered with another more LGBTQ+-friendly county, where they could expect to meet others. Another respondent said that all adult programming at the library could address LGBTQ+ issues: "That could have a positive effect on how adults treat LGBTQ+ youth." This staffer's idea to educate the entire community is a valid way of dispelling prejudices and incorrect stereotypes.

Unfortunately, the discomfort with LGBTQ+ identities also existed in the workplace. The questionnaire highlighted problematic aspects of the library's employment culture. One respondent said, "It's hard to know how and where/when to 'come out' for a lot of us who are . . . LGBTQ+. Some education on the matter could help us step into society without so much trouble, I'd like to think, especially, if not more so than for the kids, but for the families of the people who are coming out." This comment highlights how employees must navigate their own sexuality within workplace culture and how library programming could contribute to this project, one that affects so many people in multiple contexts. The mention of family members is also salient, as it gives exigence to LGBTQ+ programming that addresses heterosexual knowledge around coming out. Such people may want to understand their child's or family member's struggles. Programming focused on LGBTQ+ identities ought to be offered to the community at large.

After gathering these stories and ideas, Nelson presented the findings to the library director in a report. She made it clear that when young people, regardless of their gender identity or sexual orientation, are bullied in school, exiled from their homes, and targeted for discrimination by government and church leaders, they are at risk. She sent this information by email with a formal report attached. She received a brief message thanking her for the results, but there was no other response. The lack of response has not deterred Nelson from continuing to change what she can at the library for LGBTQ+ patrons, particularly youth. The library has since held two programs for LGBTQ+ adults, and Nelson held Lavender Literacy in August 2021 and Living in Color in June 2022.

Discussion and Implications

When engaging in TTC, our overall goal should be to create room for all experiences and search out new practices to problematize narratives that silence or exclude, making lives better and bringing about change. Nelson used her TTC work at the library as a way to address a gap in community resources. Using TTC to address issues from an LGBTQ+ perspective is particularly salient, as such work can bring narratives to the forefront, change institutions that silence such voices, and address problems at work that might seem insurmountable, given resistance from supervisors or the community.

Nelson used several techniques for addressing the silencing of LGBTQ+ identities. First, she identified the problem and sought to understand it. This required that she engage in research, and while research is often seen as a formalized endeavor meant to be published in an academic journal, Nelson used research for two reasons: (1) to understand the problem more fully and (2) to pass information up the hierarchy to effect change. While her TTC didn't necessarily result in large-scale change, it did illuminate the problem and highlight LGBTQ+ voices.

Second, Nelson engaged in report writing. We know that reports of work make visible the value and quantity of work (Haller, 1997). In this case study, the report codified the information that Nelson used to understand the problem. The report makes her work visible, concrete, and accessible. While the director's response did not invite change, and more TTC could be used to follow up on that conversation, the report still exists. The director can refer to it again and again if she so chooses, and Nelson memorializes the voices of her colleagues as important and necessary to the health of the workplace and the community. Nelson also has the option of resending the report, sending the report to other administrators, or publicizing the report.

The third technique used in Nelson's TTC is that of persistence. Despite the data suggesting that LGBTQ+ programs aren't needed or are discouraged, she continued with her plans for Lavender Literacy. Nelson worked out a program with the assistant manager that focused on Advocacy as Self-Care and includes community partners such as the local Pride celebration, youth outreach, and university LGBTQ+ organizations. Nelson and the assistant branch manager also put together goodie bags for attendees that contained a book about LGBTQ+ identities, advocacy, and self-care resources, along with fun items since the program is aimed at youth ages sixteen and older. Such programming will continue to keep the conversation going and continue to draw attention to the problem.

While the director did not change anything in response to the TTC created through Nelson's research, LGBTQ+ programming has continued, in small strides, at this library. The administration has not shut down any such efforts, and perhaps Nelson's TTC will change hearts and minds. Further, the voices of the employees who participated add to the power and weight of the TTC that Nelson created.

Ethics also play an important role in understanding this case study of TTC. We see that LGBTQ+ identities are still marginalized. Mehra and Braquet (2007) noted, "The need for progressive change in people's attitudes and behaviors is essential for a community-wide acceptance of lesbians, gays, bisexuals, transgenders, and questioning . . . individuals" (p. 542). LGBTQ+ identities must not be treated as anomalies or curiosities, something to pander to once a year. Instructively, this means that LGBTQ+ programming ought to be part of mainstream conversations, as one respondent suggested teaching the community in general about risks to LGBTQ+ youth. We should challenge "the 'minoritizing' view that assumes problematic constructions of sexual identity are an issue only for lesbian, gay, bisexual, and transgendered/transsexual . . . people rather than a problem that must be addressed for the benefit of society in general" (Alexander & Wallace, 2009, p. W301). The approach of the programming is an issue to be addressed through further TTC. Nelson can use the research she gathered to bolster this claim and teach administrators how to present such identities to the community.

In fact, by paying attention to queerness, we effectuate exceptional opportunities to engage in difficult conversations regarding intimate aspects of life. This "powerfully connects the experience of difference both to literacy and political efficacy in that our ability to speak about our lives impacts our sense of freedom to participate in our society" (Alexander & Wallace, 2009, p. W304). We contend that through actively grappling with the argument that LGBTQ+ identities need to be addressed for the benefit of society, it is possible to achieve visibility critical in advancing political emancipation and social legitimacy. The library is just one place to do so, and TTC plays an important role in addressing these gaps in many sorts of institutions.

Another ethical consideration is that of where and when employees can "come out." Not only did Nelson learn about LGBTQ+ programming and the attitudes surrounding it, but she learned about her fellow employees and their struggles. TTC led to uncovering further problems within the library, one of which is that LGBTQ+ library employees regularly navigate Cox's (2019) working-closet concept. They face a shift "from hiding one's identity in the workplace to balancing the mental load of deciding when and with whom to share" (Minei et al., 2020, p. 2). The required annual staff diversity training should specifically include LGBTQ+ concerns, which are overlooked based on the results of Nelson's questionnaire. Indeed, such focus would provide a more welcome and open work environment for LGBTQ+ staff, with spillover benefits for patrons. These benefits include a willingness on the part of library employees to be open in their sexual and gender identities with patrons who may be nervous or reluctant about requesting help in finding queer materials. It would also engender a more accepting, understanding, and positive community in general for LGBTQ+ youth that would hopefully lower the number of suicides and bullying.

This case study raises questions about TTC. What happens when TTC seems to fail and the maneuvering isn't recognized, appreciated, or used? How can tactical technical communicators make concerns more visible? Why do administrations resist such work, even when the results speak to the need for change? Nelson's experiences in the case study do not answer these questions fully. But because her work is ongoing, and she continues to use tactical moves to ensure that LGBTQ+ programming occurs, we can observe that TTC may be slow moving, especially when linked to changing organizational cultures. Not only is LGBTQ+ programming slow to happen, but the library culture of silence surrounding LGBTQ+ identities needs to change. This takes time and patience, something technical communicators in India have learned. Practitioners there have engaged in observing their organizational cultures, "practicing patience, and even serving the best interests of SMEs or organizations without clear payoff. Although these options were frustrating, some practitioners reported them as their only viable way forward" (Matheson & Petersen, 2020, p. 8).

In terms of how to make TTC visible after it is ignored, technical communicators must persist. Their first attempts at TTC may not be recognized. We suggest that more maneuvering is called for and that the artifacts of TTC could be shared with other administrators, presented at a staff meeting, revised with new research, or made public. Technical communicators may need to create opportunities to share their TTC.

Conclusion

The prevailing attitude has been that "libraries are neutral spaces," which is in direct conflict with the ALA's Social Responsibilities Round Table established in 1969 (Adler, 2015, p. 485). Such a message is detrimental, as librarians can and should be community activists, and they can advocate for users, who run the gamut of social, ethnic, racial, gender, and sexual minorities. One way of advocating for users is by engaging in TPC work tactically. Just as celebrating and acknowledging other minority populations during Black History Month or Women's History Month are entirely inadequate, "so must this conclusion be reached for the social inclusion of LGBTQ and gender variant" community (Robinson, 2016, p. 172).

Advocacy and activism within workplaces can be accomplished through TTC, as it is an important way of addressing social problems through research and ethical considerations. TTC may not always yield desired results, but its use makes problems and their solutions visible, especially when codified in a written report. Activist work is enhanced by the use of TTC, which, in this case, reveals the values of the organization and the voices of employees. When organizations are uninterested in the results of TTC, the work is still documented, available, and visible. Such work can be added to or pointed to in kairotic moments. Advocating for LGBTQ+ resources at the library is one example of how TTC can play a role in community and workplace activism.

References

Adler, M. A. (2015). "Let's not homosexualize the library stacks": Liberating gays in the library catalog. *Journal of the History of Sexuality, 24*(3), 478–507. https://doi.org/10.7560/JHS24306

Alexander, J., & Wallace, D. (2009). The queer turn in composition studies: Reviewing and assessing an emerging scholarship. *College Composition and Communication, 61*(1), W300–W320.

Brumberg, J. J. (1998). *The body project: An intimate history of American girls.* Vintage.

Cox, M. B. (2019). Working closets: Mapping queer professional discourses and why professional communication studies need queer rhetorics. *Journal of Business and Technical Communication, 33*(1), 1–25. https://doi.org/10.1177/1050651918798691

de Certeau, M. (1984). *The practice of everyday life.* University of California Press.

Ding, H. (2009). Rhetorics of alternative media in an emerging epidemic: SARS, censorship, and extra-institutional risk communication. *Technical Communication Quarterly, 18*(4), 327–350. https://doi.org/10.1080/10572250903149548

Ellis, C., Adams, T. E., & Bochner, A. P. (2011). *Autoethnography: an overview. Historical social research/Historische sozialforschung, 36*(4) 273–290.

Faber, B. D. (2002). *Community action and organizational change: Image, narrative, identity.* Southern Illinois University Press.

Feenberg, A. (2002). *Transforming technology: A critical theory revisited*. Oxford University Press.

Haller, C. R. (1997). Revaluing women's work: Report writing in the North Carolina canning clubs, 1912–1916. *Technical Communication Quarterly, 6*(3), 281–292. https://doi.org/10.1207/s15427625tcq0603_4

Holladay, D. (2017). Classified conversations: Psychiatry and tactical technical communication in online spaces. *Technical Communication Quarterly, 26*(1), 8–24. https://doi.org/10.1080/10572252.2016.1257744

Hughes-Hassell, S., Overberg, E., & Harris, S. (2013). Lesbian, gay, bisexual, transgender, and questioning (LGBTQ+)-themed literature for teens: Are school libraries providing adequate collections? *School Library Research, 16*, 1–18.

IBIS. (2019). *Complete health indicator report of suicide*. Public Health Indicator Based Information System. https://ibis.health.utah.gov/ibisph-view/indicator/complete_profile/SuicDth.html

Kimball, M. A. (2006). Cars, culture, and tactical technical communication. *Technical Communication Quarterly, 15*(1), 67–86. https://doi.org/10.1207/s15427625tcq1501_6

Kimball, M. A. (2017). Tactical technical communication. *Technical Communication Quarterly, 26*(1), 1–7. https://doi.org/10.1080/10572252.2017.1259428

Matheson, B., & Petersen, E. J. (2020). Tactics for professional legitimacy: An apparent feminist analysis of Indian women's experiences in technical communication. *Technical Communication Quarterly, 29*(4), 376–391. https://doi.org/10.1080/10572252.2019.1659860

McCutcheon, J. M., & Morrison, M. A. (2021). Beyond the superordinate categories of "gay men" and "lesbian women": Identification of gay and lesbian subgroups. *Journal of Homosexuality, 68*(1), 112–137. https://doi.org/10.1080/00918369.2019.1627129

Mehra, B., & Braquet, D. (2007). Library and information science professionals as community action researchers in an academic setting: Top ten directions to further institutional change for people of diverse sexual orientations and gender identities. *Library Trends, 56*(2), 542–565. https://doi.org/10.1353/lib.2008.0005

Mehra, B., & Braquet, D. (2011). Progressive LGBTQ reference: Coming out in the 21st century. *Reference Services Review, 39*(3), 401–422.

Minei, E. M., Hastings, S. O., & Warren, S. (2020). LGBTQ+ sensemaking: The mental load of identifying workplace allies. *International Journal of Business Communication, 60*(3), 823–843. https://doi.org/10.1177/2329488420965667

Montague, R. A. (2015, August 15). *Leading and transforming LGBTQ library services* [Conference Presentation]. IFLA Conference Proceedings, Cape Town, South Africa.

Peters, B., & Swanson, D. (2004). Queering the conflicts: What LGBT students can teach us in the classroom and online. *Computers and Composition, 21*(3), 295–313. https://doi.org/10.1016/j.compcom.2004.05.004

Petersen, E. J. (2018). Female practitioners' advocacy and activism: Using technical communication for social justice goals. In Godwin Y. Agboka & Natalia Matveeva (Eds.), *Citizenship and advocacy in technical communication: Scholarly and pedagogical perspectives* (pp. 3–22). Routledge.

Robinson, T. (2016). Overcoming social exclusion in public library services to LGBTQ and gender variant youth. *Public Library Quarterly, 35*(3), 161–174.

Ryan, C., Russell, S. T., Huebner, D., Diaz, R., & Sanchez, J. (2010). Family acceptance in adolescence and the health of LGBT young adults. *Journal of Child and Adolescent Psychiatric Nursing, 23*(4), 205–213. https://doi.org/10.1080/01616846.2016.1210439

Sang, J. M., Louth-Marquez, W., Henderson, E. R., Egan, J. E., Chugani, C. D., Hunter, S. C., Espelage, D., Friedman, M.S., & Coulter, R. W. (2020). "It's not okay for you to call me that": How sexual and gender minority youth cope with bullying victimization. *Journal of Homosexuality, 3*, 1–20. https://doi.org/10.1080/00918369.2020.1826831

Sarat-St. Peter, H. A. (2017). "Make a bomb in the kitchen of your mom": Jihadist tactical technical communication and the everyday practice of cooking. *Technical Communication Quarterly, 26*(1), 76–91. https://doi.org/10.1080/10572252.2016.1275862

Stringer-Stanback, K. (2011). Young adult lesbian, gay, bisexual, transgender, and questioning (LGBTQ) non-fiction collections and countywide anti-discrimination policies. *Urban Library Journal, 17*(1), 1–27.

Chapter Three

Exposing the "Actively Enforced" Policy

Tactical Technical Disruptake for Rhetorical Deinstitutionalization

Walker P. Smith

With megasized campuses around the world, the evangelical Hillsong Church boasts a weekly attendance of 150,000 people. In 2014, the *New York Times* reported the church had shifted their "tone" on gay marriage, citing lead pastor Brian Houston's effort to "stay relevant as a church" (Paulson, 2014, para. 5). The church released (and later deleted) a statement to "clarify" their position: "We are an inclusive Christian church that loves, values, and welcomes all people, regardless of their background, ethnicity, beliefs, values, or personal identity" (Church Clarity, Feb. 2019, para. 3). Why, then, was their music director, Josh Canfield, fired from his leadership role after he publicly came out as a gay Christian? (Church Clarity, Mar. 2019, para. 2). Their official position may be found only through in-depth research into their media archives: "Hillsong Church welcomes ALL people but does not affirm all lifestyles," later followed by, "So if you are gay, are you welcome at Hillsong Church? Of course!" (Houston, 2015, paras. 3 & 7). While the tone might seem accepting, this statement actually implies that all nonheterosexual attendees of Hillsong will be instructed to remain celibate until they participate in some form of conversion therapy and/or a heterosexual marriage.

Misleading bait-and-switch rhetorics such as Hillsong's prompted the volunteer-run organization Church Clarity to form in 2017. Its mission is to find and score the "actively enforced" LGBTQ policies of Christian churches so that users of their website can make informed decisions about the spiritual spaces they want to inhabit. Actively enforced policies are not what churches purport to believe; they are "what churches do" (Church Clarity, "Our Scoring," n.d., para. 4). For example, rather than asking Hillsong members if they believe they are welcoming, what we should be asking them are specific questions about practice, such as the following: Will you officiate a same-sex wedding? Are transgender people eligible to serve in leadership and preach? Today, volunteers continue to expand the database of church scores, amidst an aggressive political climate in which US state legislatures propose and pass legislation that permits religious discrimination and seeks to "eradicate trans youth and trans lives" (Strangio, 2022, para. 1).

Church Clarity's work may also be understood as what Dryer (2016) calls genre "disruptake," which he defines as "uptake affordances that deliberately create interficiencies, misfires, and occasions for second-guessing that could thwart automaticity-based *uptake enactments*" (p. 70). Their volunteers deploy extensive research methods to intervene in uptake processes ("the interconnections, translations, and pathways between genres" [Reiff & Bawarshi, 2016, p. 3]) as they completely revise and republish the policies to be more honest and accurate. Such interventions lead me to ask: How does disruptake function as a form of tactical technical communication, and what might be its rhetorical impacts on institutionalized genres? In exploring this question, I share interviews with two users of the Church Clarity website to understand how they interpret disruptake and to capture the affective consequences of initiating the rhetorical processes of deinstitutionalization ("the process by which deeply entrenched practices give way to new innovations" [Ahmadjian & Robinson, 2001, p. 622]).

Church Clarity as Tactical Technical Communication

Many Christian churches already hire technical writers for their daily operations, some of whom actively participate in the Society of Technical Communication and regularly present at conferences (Johnson, 2009). In their own words, the technical writers at the Church of Jesus Christ of Latter-Day Saints are asked to "create a variety of help materials, including software manuals, online help, quick reference guides, video tutorials, interface text, release notes, web content, and product communications" (Johnson, 2009, n.p.). One aspect of their work is to craft websites as public-facing texts that welcome new members, guide current members to become more involved in church activities, and instruct all users on how the church plans to interpret biblical scripture and enact those beliefs as institutional doctrine.

While the fields of technical and professional communication have traditionally described business-driven workplaces, scholars are continually revising our dominant narratives beyond the "narrow context" of "business environments," which are "insufficient for framing the role and influence of our work" (Walton, et al., 2016, p. 86) and instead privilege "social justice and inclusivity" as on par with "workplace problem solving" (Jones, et al., 2016, pp. 212–13). This chapter contributes to their turn by recognizing churches as crucial sites of institutional messaging. Churches wield significant meaning-making power over their congregations by crafting technical rhetorics that project certain approaches to biblical interpretation as instructional guides for how to live morally. Thus, the field would benefit from learning more about how users' interactions with church materials impact their daily lives.

Church Clarity "[works] outside" of these churches to intervene in their exchanges by sharing extra-institutional risk communication that reduces the amount of power churches hold over their members (Kimball, 2017, p. 1). Their user-driven, volunteer-built scoring systems, as well as the ongoing labor of maintaining, updating, and expanding their online database of church scores, requires a series of skills: the application of subject-specific knowledge about evangelical belief and practice; extensive training into their organization's research methodology and what counts as a credible source, which was curated through community-driven criteria; recomposition of complex and obscured multimodal messages written by institutions into clear, accessible summaries according to a consistent style guide; and collaboration with a team of volunteers who use a streamlined process that verifies every score as valid before it's published. Their mission to score every church, "rate the validity," and "provide valuable metrics on the product and its use" is a massive undertaking that applies many of the skills often taught in technical-communication courses (Kimball, 2017, p. 2). Though it lacks a formal discussion-board space that is typical for tactical technical communicators (Holladay, 2017), Church Clarity is still "creating space for marginalized users' expertise to be recognized as legitimate" by inviting them to share their narrative experiences, and in turn helping others to make more "informed decisions" (Jones, et al., 2016, p. 218). By filtering churches' technical communication through community-based criteria, it challenges who has the power to write policies and enact practices, offering them an "extra-institutional [channel] to challenge and contradict official media and to communicate imminent risks to the public" (Ding, 2009, p. 329).

For this case study, I interpret Church Clarity's tactical technical communication through the lens of rhetorical genre studies (RGS). To safely navigate spiritual spaces, LGBTQ+ people need access to the stock of genre knowledge, or what RGS scholars understand to be the situated practices that reproduce social structures and are the basis for successful social interactivity (Berkenkotter & Huckin, 1993). The practices that actively enforce church policies—for example, how a church leader responds to a member revealing their LGBTQ+ identity—becomes institutionalized as a genre through consistent uptake, which works "to justify and objectify" churches' right to make meaning in the lives of its members (Britt, 2006, p. 137). Once a genre is institutionalized for a particular public, it "[works] in the background to both draw upon and shape what we perceive as common sense" (Britt, 2006, p. 134). Church Clarity targets those "occluded genres" that are hidden from some members "by a veil of confidentiality" in an effort to rebalance the power dynamics at work (Swales, 1996, p. 46). Although churches purport to be a safe space for discussion and debate over biblical interpretation and learning more about those occluded genres (e.g., Hillsong's invitation to participate in an "ongoing conversation"), requests to do so may often result in harmful consequences for LGBTQ+ members (e.g., Hillsong's removal of Josh Canfield from a leadership position). Through sermons, small-group discussions, and one-on-one counseling, church leaders dictate how members may express their gender and sexuality and how other members should respond to this expression—enacting uptake affordances and constraints that "facilitate particular uses or deter particular activities" (Dryer, 2016, p. 65).

Genre uptakes, whether successful or thwarted, "all help maintain, modify, and destabilize cultural *institutions*" (Dryer, 2016, p. 65, emphasis added). In other words, one of the consequences of Church Clarity's efforts toward disruptake (disrupting uptake) may be to begin the processes of deinstitutionalization for its users, which is often initiated through "the authoring of texts that problematize existing practices" (Maguire & Hardy, 2009, p. 168). Though Clarity's immediate goal is to reduce harm, in the process of harm reduction, members may cause some institutions to individually unravel for users, to make less sense than they did before. In the next section, I will describe key moments of disruptake from three interviews with users of the Church Clarity website and explore their affective consequences or the beginning stages of deinstitutionalization.

"I'm Just Like, What Are You Hiding?": How Users Interact With Church Websites

I conducted interviews with users and not volunteers for one key reason: while volunteers are able to demonstrate how they enact tactical technical communication to meet their goals of deinstitutionalizing uptakes of church policies, only users of the website can actually validate whether/how their goals were effectively met. Additionally, users are able to report the aftereffects of experiencing a disruptake of a previously trusted genre, which the study showed can sometimes lead to unpredictable affective consequences. I located participants by sharing an IRB-approved invite in two Facebook groups devoted to LGBTQ+-identified Christians and one Facebook group for "exvangelicals," or people who have left the evangelical church or are considering leaving. These groups were chosen because they each already had archived posts sharing and discussing Church Clarity's services.

Below, I share interviews with two participants because they each brought uniquely diverse histories and feelings in relation to Christianity: Sarah is a bisexual mother who attends a conservative, evangelical church and remains closeted, and Jill is a gender-nonconforming lesbian who identifies as a progressive Christian "pastor's kid" but doesn't regularly attend any church. For Jill, much of the typical evangelical teachings on gender and sexuality had already been deinstitutionalized prior to the interviews, as they already displayed confidence in their identities and reconciliation with traditional transphobic and homophobic interpretations of scripture. Sarah, on the other hand, felt conflicted about and hid her sexual activity with women, while publicly supporting her gay relative who attends church with her.

My methodological approach to the interviews was grounded in queer usability, which provides "supplementary ethical frameworks" (Edenfield et al., 2019, p. 181) for scholars in the field to reduce harm and "[center] marginalized users and their anticipated needs" (Ramler, 2020, p. 1, 10). Each interview was divided into two phases: interviewees were asked to interact with four church websites and then asked to interact with the Church Clarity scores corresponding with those four churches. This structure of moving from church websites to Church Clarity scores helped to identify what exactly the disruptakes were disrupting. Three of the churches were selected using the three categories provided by Outreach100.com, an annual evangelical report that gathers data on church growth in the United States: fastest growing, most reproductive (expanding to more locations), and largest in size. The churches chosen from these categories were the highest-ranking churches that had also already been scored by Church Clarity. The fourth church was selected because it is scored as LGBTQ+ affirming by Church Clarity and because it poses a distinct contrast in digital presence in comparison to the other three. In the first phases, users were asked to think aloud as they browsed each church's website and talk through the process of answering the following questions: (1) Would you attend this church? (2) Do you think they're affirming? In the second phases, users were asked to think aloud as they tried to read and make sense of Church Clarity's score and its reasoning behind the scores, answering the following questions: (3) Do you agree with Church Clarity's rewritten policy, score, and reasoning? (4) Do you think publishing this score is effective for positive institutional change?

Both participants struggled the most with question 2: "Do you think they're affirming?" This question caused anxious deliberation, second-guessing of previously reliable knowledge, and even outright frustration. While they always had immediate reactions to the aesthetics of the church website's homepage, question 2 led participants into deep browsing dives, and rarely were they absolutely confident that they had come to the right conclusion. However, Jill was still able to determine correctly whether all four churches were affirming, even when they were admittedly guessing, while Sarah failed to identify one of the four churches as nonaffirming. Additionally, when Sarah could successfully identify a church's affirming or nonaffirming status, her justification was often very disconnected from their actual stances on gender and sexuality. While Church Clarity's services would have saved both of the participants a lot of time, effort, and risk, only Sarah would have been potentially exposed to harm without Church Clarity's guidance.

Interestingly, though, even when participants identified the status correctly, they were almost always operating with different criteria than Church Clarity uses. While Church Clarity volunteers ask questions such as, "Would this church officiate a same-sex wedding?" and "Are all LGBTQ+ people eligible to serve in leadership and preach?," participants relied on the following criteria to make their decisions:

In all, participants relied much more heavily on criteria that gave them clues that might predict what a church's policy is, but only about 25% of the time did they actually find enough proof that Church Clarity volunteers

Table 3.1. User-Generated Criteria for Church Webtext Evaluation

User-Generated Criteria	Sarah (13 total instances of using criteria)	Jill (25 total instances of using criteria)
Age of members (in images)	1 mention	
Ambiguously invitational rhetoric (e.g., "you are welcomed here")		2 mentions
A clear, accessible "open and affirming" statement		2 mentions
Community service (local, and distinct from evangelical missions)		3 mentions
Defining terms		1 mention
Denominational affiliation		1 mention
Design quality	1 mention	2 mentions
Encourages debate or questioning of scripture	1 mention	2 mentions
Geographic location	1 mention	
Masking protocols	2 mentions	2 mentions
Name of church		2 mentions
Racial and ethnic diversity in images		1 mention
Scriptural interpretation		2 mentions
Size	4 mentions	2 mentions
Theatrics (sound and stage production)	3 mentions	1 mention
Written references to diversity and inclusion		2 mentions

Source: Created by the author.

would qualify as publishable evidence. In particular, Sarah found the least amount of evidence overall, little more than half of the other participants. While she tended to be more focused on the church's size and theatrics, she missed much of what Jill found most useful: the ratio of ambiguously invitational rhetoric to clear statements of belief. More "welcoming" and "accepting" and a lack of prominent affirming declarations often led to the feeling that "they're being purposefully vague." One spoke directly to this feeling as they browsed Radiant Church's website, which, other than its inexplicable "heteronormative vibe," left them with much suspicion: "I'm sure they meant that to be open and welcoming, but to me, I'm just like, what are you hiding? What are you hiding?"

In contrast, Sarah found Radiant's website to be "aesthetically pleasing" and reported positive associations with the belief categories that sparked suspicion in others. Ultimately, she correctly predicted that the church would be nonaffirming of LGBTQ+ members, but not because of the lack of clear belief statements. Instead, she focused on the images they shared on the main page that didn't show enough evidence of a "ton of younger people." For example, in one image, a large audience listens to a casually clad speaker at a minimalist podium. The words "We want to help you discover the purpose God has for your life" superimposed over the image limits a full assessment of the audience's ages. Even so few appear to be under 20.

Likely influenced by a false teaching commonly found in evangelical churches' LGBTQ+ policies—that out LGBTQ+ Christians have emerged more recently as a contemporary phenomenon—Sarah associates affirming theologies with millennials: "I feel like more millennials are ushering in that [. . .] paradigm shift [. . .] of being more welcoming and accepting of others. I feel like the majority of the crowd looks to be a little bit older." The first phase of interviews showed that users like Sarah are at significant risk of repeated religious trauma and need more helpful and protective tools when navigating church websites and their policies. Bait-and-switch rhetorics like those displayed in Radiant's materials provide the appearance that LGBTQ+ members, along with all others, will be welcomed, accepted, and loved as they are. However, we know from their other policies that this acceptance will require the member to attempt to deny, change, and/or hide their LGBTQ+ identity.

"Like a Mining Process": How Users Interact with Church Clarity Disruptakes

In the second phase of the interviews, as Church Clarity's scores were revealed for the four churches in the study, both participants had affective reactions. Even though both participants later failed to understand Church Clarity's scoring system without my intervening guidance, the disruptakes were still an effective kickstart to initiate microlevel processes of deinstitutionalizing certain genres.

Jill's Relief (City Church, Oklahoma)

As a self-identified, progressive Christian, Jill doesn't currently attend a church for many reasons, one of which is the amount of research required in finding one. When looking for one to attend, they demand "a direct overture": "I don't need random [LGBTQ+ pride] flags. I don't need a whole page just left to it, but I want to see more than a two-second slide that passes by." In the past, they spent a significant amount of time scrolling through endless pages of sermon archives and skipping around each audio file, searching for the right clip that might indicate their LGBTQ+ policy, ultimately "to see if, A, I like the way that the preacher preaches and, B, if it was absolutely offensive." Jill eventually started reading user-submitted reviews of churches on Google's business pages, where they say they can often find an indication that the church is transphobic or homophobic. To meet their standards, though, they would want to see a reviewer specifically mention that the church is involved in some kind of local mission efforts supporting LGBTQ+ youth.

Upon opening City Church's "fancy website" in the first phase of the interview, Jill was mesmerized by the design of what they immediately termed a "megachurch": "The photos, the amount of people, the double doors. [. . .] I'm overwhelmed by this website, first and foremost, just the amount of moving pictures." This "shifting of things" left them feeling somewhat lost, so they clicked on a "three lines" icon hoping for a menu. The menu led them to a church planting page (which received a quick "I don't like that phrase" for its "corporate speak"), a missions page, and an about page. Despite the abundance of information on those last two pages, Jill felt confused: Do they practice *good* missions ("meeting the needs") or *bad* missions ("conversion trips")? When they say they are "committed to be doers," does it mean they will "idly judge by the standards of the Bible" or actually "practice what they preach" and help those in "the margins of society?" With "so many things on this page," Jill was frustrated that none of the content indicates a clear stance or policy. Ultimately, they guessed that the church is nonaffirming because of its nod to "cultural diversity" and "different races, ethnicities, and heritages," without making any similar mentions of gender and sexual diversity. Still, though, they were left with uncertainty, immediately asking me after making that decision whether they were being "too picky."

Later, in the second phase of the interview, in which they were asked to read and comment on Church Clarity's score for City Church, Jill's frustration and lack of assurance was quickly relieved. Within seconds, they first opened the provided link to the 20:16 mark on a 2018 sermon titled "Kingdom Life—Sexuality & Lust." After watching the short clip, they described the ease of access, stating, "Yeah. I watched it. I watched that one line. Wow. That's enough. Well, I think that that is . . . I love that. I love that not only is it sending me to the sermon, but it's sending me right to the line. That is so helpful. Jeez. That's so upsetting, too, because it does seem like they were doing a lot of good community things, but yeah."

In this type of disruptake, Jill was overwhelmed as a user and knew that it will take them potentially long hours of research to locate a Google review or other buried piece of evidence that might or might not accurately indicate the church's LGBTQ+ policy. Church Clarity has essentially mapped out those pathways between genres for users—or how the church intends to respond to gender and sexual diversity in its congregation—and thus disrupts the church's intended rhetorical strategies of pretending to welcome new LGBTQ+ members before rejecting them.

This act of disruptake accelerated Jill's own deinstitutionalization, in which old practices are replaced with or improved by new innovations, by elevating their tactics from long sessions of reading Google reviews and scrolling sermon archives to a faster, more accurate search of Church Clarity's website. In other TTC projects, similar reactions of relief may be achieved by collapsing complex genre systems to make them more navigable and shortening the number of steps it takes to locate occluded knowledge. The more streamlined user experience may fast track projects working toward deinstitutionalization by reducing the amount of effort that users must input in order to access the information.

Sarah's Revelation (Eagle Brook Church, Minnesota)

By the time Sarah reached the final church in the first phase of the interview, she had developed a good sense of what criteria she *wasn't* looking for in a church website: size, and what she called "theatrics." The larger the crowds shown in images, the more likely it was to be nonaffirming from Sarah's perspective. If the sound and stage production seemed expensive, then she would conclude that it is more likely focused on expansion and not community service: "I'm not a fan of, like I said, theatrics. I don't think that it's so cool that the children's church has all this air hockey and stuff like that. I feel like that's not really the purpose of church." When she encountered the website of the only affirming church in the study, she valued that it didn't seem to be focused on expansion: "I love that church, and I'm glad that I got it right because I feel like, I mean, I could just tell by the way it was set up. It seemed like they don't give a freaking shit about numbers."

The task of determining if churches were affirming became the most challenging for Sarah when size and theatrics didn't seem to match up with some of the other practices shown on the church website. For example, the stage at Eaglebrook Brook Church in Minnesota appears more in keeping with a professional concert venue. At least eight musicians perform spread out on a vast stage with under-the-floor lighting and what appear to be special lighting effects. This expensive-looking stage coupled with the large crowds and 11 different locations suggested to her that the church would be nonaffirming

However, many of the other images showed members wearing face masks, likely to prevent the transmission of the COVID-19 virus. She explained that her own home church—a nonaffirming one that has nearly 40 locations spanning almost 10 states—does not require masks for in-person attendance. When I suggested that Centerville, Minnesota, could have different local ordinances regarding mask usage, she insisted that a conservative evangelical church likely wouldn't comply, adding: "[Eagle Brook members] have all these social-distancing things on their floors in the sanctuary, but like I said, because of that I would assume that they would probably be affirming." Thus, conflicting political clues further obscured the church's LGBTQ+ policies for her.

In the second phase of the interview, Sarah was upset to learn that the church was nonaffirming and that Church Clarity had discovered this knowledge in two ways: (1) a 2013 sermon from the archives, at the 20:55 and 21:42 marks, and (2) its name was listed in the Converge General Baptist Conference denomination's church-locator search engine. Even though Converge was nowhere referenced on Eagle Brook's website at the time, the denomination listed its congregation as a member, and Converge listed its many nonaffirming policies on its denomination's website. Even though Sarah had previously attended and worked for at least one church that hid its denominational affiliation from most of its members, she hadn't yet fully realized that this move was an intentional one—a rhetorical effort to suppress the spread of its LGBTQ+ policies. Sarah was very angered by Eagle Brook's decision to hide this information from the website: "Because it's not shown, you either have to go and find out for yourself or you have to find out in a really awkward and rejecting way by hearing the main man on the stand basically bash who you are. I don't like that." What she didn't mention is that, in this hypothetical scenario, "finding out" could take many years. A new member might not encounter that information in a sermon since Eagle Brook's sermons on the topics of gender and sexuality only appeared in the archives once or twice a decade.

Sarah's anger and frustration at this move led her further into a revelation about how genres function rhetorically in nonaffirming churches. She continued to talk her way through a church's reasons for this move, becoming more upset as she spoke and connecting it to her previous criteria of church size and theatrics:

> Yeah, I think it's weird that they wouldn't share that, but it goes back to the whole purpose of why they're not, because they want some random Joe off the street coming even if their basic doctrines does not line up with the church's, because it's just a seat that's being filled. [. . .] I mean, you might as well put that your mission is to fill as many seats as possible and fill your pockets because that's really what you're doing.

This new connection disrupted previously uninterrogated logics for her, as she began to think about how nonaffirming believers should also want the policies to be clarified: "I feel like it's weird that churches are moving away from that, even in a mission statement." During the interview itself, Church Clarity's scoring protocols revealed to Sarah how the active enforcement of church policies can be obscured from its members and what powerful leverages that may provide its leaders in the process.

Because Church Clarity volunteers modeled how to locate and extrapolate a church's LGBTQ+ policies within a tangled and complex web of data and linked websites, this revelation strengthened Sarah's analytical skills by teaching her how to navigate those websites and policies in the future. The revelation also laid the ground for further deinstitutionalization, as she now has a more critical orientation to institutionalized genres that publicly describe church policies. Sarah's revelation reminds us that projects similar to Church Clarity should build opportunities for comparison and reflection all throughout their materials. Sarah would not have compared her own church's policy and practices to Eagle Brook's unless I had explicitly asked her to. In order for deinstitutionalization to begin, it's important to specifically prompt users to make connections between new information and their past experiences.

Conclusion

In these two interviews, relief and revelation were productive disruptake experiences in which users discovered how to save time and effort, how to strengthen their analytical and navigational skills, and how to adopt a more critical orientation toward institutional messaging. Disrupting the habitual operations (uptakes) of institutionalized genres proved to be an effective form of tactical technical communication because users were able to work outside of the institution to exchange information and gain new knowledge about the potential risks they faced.

Additionally, each disruptake contributed to the deinstitutionalization of certain genres, "the process by which the legitimacy of an established or institutionalized organizational practice erodes or discontinues" (Oliver, 1992, p. 564). Church Clarity optimistically hopes that initiating this process for its users will ultimately contribute to larger-scale changes in evangelical practice. If enough stakeholders no longer support the continued use of an institutionalized bait-and-switch genre, then ideally church leaders will no longer have the "ability" to "continually recreate an institutionalized organizational activity" without criticism or even delegitimization (Oliver, 1992, p. 564). However, I would argue that rhetorical deinstitutionalization should not be understood as a singular end goal, such as all church leaders abandoning the use of bait-and-switch rhetorics, which can be naive and unlikely to actualize.

Instead, I understand rhetorical deinstitutionalization as a set of highly contested, often shifting responses to institutionalized genres. Users frequently work alone or together to circumvent, foil, and overcome certain uptake affordances, and their collective efforts may, or may not, gradually chip away at the power of institutions to make meaning in their lives. Over time, certain tactics may increase the likelihood of deinstitutionalization to succeed and to spread, such as composing clear and accessible technical guides and publishing them online, but it is not inherently a guarantee that every tactic will result in any amount of deinstitutionalization.

While volunteers' rewritten policies were clear and provided enough links to outside resources for users to successfully read and apply them, the score rating itself occasionally caused confusion. After I shared Church Clarity's technical guide for understanding scores, though, users quickly understood and agreed with the label. But the lack of quick access to that technical guide on the score page temporarily resulted in a reduction of Church Clarity's ethos, weakening their trust in or understanding of the scoring system. In pushing for disruptake, tactical technical communicators must provide swift access to their technical guides. In this case, reducing the amount of digging users need to perform is ideal for limiting the unpredictable disorder that can result from disruptakes, and queer usability research is especially useful in identifying where breakdowns occur in content navigation. However, simple design may need to be balanced with the delivery of technical knowledge. Deinstitutionalization can much more safely be controlled by directly helping users navigate their new knowledge, and guides will need to gradually and gently replace old and existing knowledge in order to effectively implement tactical technical disruptake in other contexts.

References

Ahmadjian, C., & Robinson, P. (2001). Safety in numbers: Downsizing and the deinstitutionalization of permanent employment in Japan. *Administrative Science Quarterly, 46*(4), 622–654.

Berkenkotter, C., & Huckin, T. N. (1993). Rethinking genre from a sociocognitive perspective. *Written Communication, 10*(4), 475–509.

Britt, E. (2006). The rhetorical work of institutions. In J. B. Scott, K. V. Wills, & B. Longo (Eds.), *Critical power tools: Technical communication and cultural studies* (pp. 133–150). State University of New York Press.

Christianity Today. (Eds.) (2022, March 22). *Hillsong Church founder Brian Houston resigns.* Christianity Today. https://www.christianitytoday.com/news/2022/march/hillsong-brian-houston-text-hotel-complaint-board-investiga.html

Church Clarity. (n.d.). *Our Scoring.* Church Clarity. https://www.churchclarity.org/about

Church Clarity. (2019, February 8). *Church Clarity profile: Hillsong NYC.* Church Clarity. https://www.churchclarity.org/updates/church-clarity-responds-to-hillsongs-statement-on-inclusion

Church Clarity. (2019, March). *Gay, ex-Hillsong leader speaks about experience with unclear policy.* Church Clarity. https://www.churchclarity.org/updates/former-gay-worship-leader-for-hillsong-nyc-speaks-on-experience-with-unclear-policy

Church Clarity. (2020, June 2). *Living Faith Covenant Church.* Church Clarity. https://www.churchclarity.org/church/living-faith-covenant-church-3217

Ding, H. (2009). Rhetorics of alternative media in an emerging epidemic: SARS, censorship, and extra-institutional risk communication. *Technical Communication Quarterly, 18*(4), 327–350. https://doi.org/10.1080/10572250903149548

Dryer, D. B. (2016). Disambiguating uptake: Toward a tactical research agenda on citizens' writing. In M. Reiff & A. Bawarshi (Eds.), *Genre and the performance of publics* (pp. 60–79). Utah State University Press.

Edenfield, A., Holmes, S., & Colton, J. (2019). Queering tactical technical communication: DIY HRT. *Technical Communication Quarterly, 28*(3), 177–191. https://doi.org/10.1080/10572252.2019.1607906

Holladay, D. (2017). Classified conversations: Psychiatry and tactical technical communication in online spaces. *Technical Communication Quarterly, 26*(1), 8–24. https://doi.org/10.1080/10572252.2016.1257744

Houston, B. (2015, August 4). *Do I love gay people?* Hillsong. https://hillsong.com/collected/blog/2015/08/do-i-love-gay-people/#.YRgA9cYpBap

Johnson, T. (2009, April 29). *Does the church really have technical writers?* The Church of Jesus Christ of Latter-Day Saints. https://tech.churchofjesuschrist.org/blog/221-tom-johnson

Jones, N., Moore, K., & Walton, R. (2016). Disrupting the past to disrupt the future: An antenarrative of technical communication. *Technical Communication Quarterly, 25*(4), 211–229. https://doi.org/10.1080/10572252.2016.1224655

Kimball, M. (2017). Tactical technical communication. *Technical Communication Quarterly, 26*(1), 1–7. https://doi.org/10.1080/10572252.2017.1259428

Maguire, S., & Hardy, C. (2009). Discourse and deinstitutionalization: The decline of DDT. *Academy of Management Journal, 52*(1), 148–178. https://doi.org/10.5465/amj.2009.36461993

Oliver, C. (1992). The antecedents of deinstitutionalization. *Organization Studies, 13*(4), 563–588. https://doi.org/10.1177/017084069201300403

Paulson, M. (2014, October 17). *Megachurch pastor signals shift in tone on gay marriage.* The New York Times. https://www.nytimes.com/2014/10/18/us/megachurch-pastor-signals-shift-in-tone-on-gay-marriage.html

Ramler, M. E. (2020). Queer usability. *Technical Communication Quarterly, 30*(4), 345–358. https://doi.org/10.1080/10572252.2020.1831614

Reiff, M., & A. Bawarshi. (2016). *Genre and the performance of publics.* Utah State University Press.

Strangio, C. Interview with A. Goodman. (2022, March 9). *Chase Strangio on the GOP's push in Florida, Texas, Idaho.* Democracy Now!. https://www.democracynow.org/2022/3/9/chase_strangio_on_anti_trans_legislation

Swales, J. (1996). Occluded genres in the academy: The case of the submission letter. In E. Ventola & A. Mauranen (Eds.), *Academic writing: Intercultural and textual issues* (pp. 45–58). John Benjamins.

Walton, R., Mays, R., & Haselkorn, M. (2016). Enacting humanitarian culture: How technical communication facilitates successful humanitarian work. *Technical Communication, 63*(2), 85–100.

Chapter Four

#WearAMaskNY Public Service Announcements

Tactical Technical Communication During a Global Pandemic

Shannon N. Sarantakos and Sara C. Doan

The State of New York has received much attention during the Coronavirus pandemic, first for its high number of total COVID cases and COVID-related deaths (Do & Frank, 2021) and later for its strict but effective measures to combat the virus's spread. In May 2020, New York State sponsored a competition inviting New Yorkers to submit their own videos for public service announcements (PSAs). This chapter examines how the New York State government officials successfully utilized tactical technical communication (TTC) to support their message about the importance of wearing a mask to prevent COVID-19's spread. This competition encouraged TTC, allowing contestants to use their own voices, the voices of the people, while simultaneously spreading the state's message. In doing so, New York gave away the power to tell its message in its own words but gained the power and support of its constituents. Users took on this task enthusiastically; over 600 New Yorkers submitted videos from May 5 to May 26, 2020. While promoting the overall message to wear a mask, contestants employed their own tactics to advocate for specific groups that mattered most to them, such as health-care workers and those living in medically underserved areas.

This competition exemplifies two recent movements in technical communication: TTC and the social-justice turn (e.g., Walton et al., 2019). With expanding digital tools, "everyone who enjoys access to the Internet is now a potential technical communicator" (Kimball, 2017, p. 1), allowing greater opportunities for underrepresented people to generate technical communication. TTC can spread health information during epidemics (Ding, 2009), promote patient understanding (Bellwoar, 2012), and make information more accessible for historically marginalized groups (Holladay, 2017). However, integrating user-generated material into institutional communication can violate users' expectations or create questionable content (Reardon et al., 2017), making organizations hesitant to involve users in the process of promoting institutional goals. Additionally, tensions between strategies and tactics during epidemics (Ding, 2009) make the context surrounding New York State's PSA contest particularly precarious. By harnessing tactics for use as strategies, however, New York invited submissions that represent a type of cocreation between the state and the people.

The "Wear a Mask" campaign thus represents an institution's successful attempt to leverage users' content to advance its goals through cocreation (Reardon et al., 2017). This chapter critically examines tactics that the five video finalists used to advance their own agendas while also supporting the state's official strategy. Some finalists challenged viewers to "prove it together" and "not be that guy," while others brought attention to groups, both broadly (essential workers, marginalized populations, immunocompromised) and narrowly ("my momma who's a health-care worker") defined. Others focused on how wearing a mask would help New Yorkers get back to normal activities such as sharing a meal and returning to work. The participation of the public in this government-sponsored contest was integral to its success, both as a competition and in disseminating health information. Like Pflugfelder (2017), we assert that public-facing genres, like public service announcements, distill technical communication strategies into plain language, transforming public-health research into action items for the lay audience: New Yorkers.

Our examination of PSA videos reveals what happens to institutions' strategies and users' tactics when messages are created and framed outside of an organization but approved and distributed from a larger organization, thus mediating the relationship between the New York State government and its citizens. This contest exemplifies "active reception" (Bellwoar, 2012, p. 325) of shifting health information, a "cocreation" (Reardon et al., 2017, p. 42) between the state and the people to give citizens agency in sharing its message in the age of social media. "Production is a combination of making and using—of strategic and tactical action—of production by owner-makers and consumption (i.e., another form of production) by consumer users" (Reardon, et al., 2017, pp. 52–53). What further complicates TTC is ethical questions around compensating, acknowledging, and coopting users' contributions. Although our analysis shifts to government-sponsored communication, we still find this idea of cocreation useful, as it offers opportunities for reciprocity in highly kairotic contexts. TTC serves as a lens for analyzing power differentials' effects on communication in health contexts (Edenfield et al., 2019), including examining the impact of professional production among these user-producers.

This chapter explores how TTC can bring organizations in power and everyday users together to advance each of their causes. While these videos allow underserved people to show their own needs within institutional boundaries, we posit that tactics do not always exist in opposition to strategies (Colton et al., 2017). Tactics do allow the manipulation of strategies for personal ends, as government institutions depend on strategies to assert their own legitimacies during this posttruth era (Walwema, 2021). This chapter asks, Do strategies and tactics need to be at odds, or can they beneficially work together? Given the social-justice turn, cooperation might serve technical communicators well by considering how professional technical communicators cultivate tactical ways of thinking, writing, and engaging underrepresented voices.

Tactical Technical Communication in Post-Postmodern Life

During our current user-centered turn, TTC reflects the tensions between institutions and individuals during the post-postmodern era of late capitalism. With the economic downturns of 2008's great recession and COVID-19, symbolic-analytic work (Wilson & Wolford, 2017) has become more fractured than ever. This context of downturn means that "technical communicators need to start viewing audiences as potential collaborators and participants rather than just users" (Reardon et al., 2017, p. 54), even while TPC as a field continues to grapple with its own identity (Henning & Bemer, 2016). Further complicating tensions between cocreation and gatekeeping are moves for decolonizing research methods (Phelps, 2021) and incorporating a feminist ethics of care (Colton et al., 2017).

Strategies and Tactics

As the purview of institutions, strategies can sometimes inhibit necessary change, preserving the status quo instead of prompting repair to systems that support ongoing harm and discrimination. Institutional checks, balances, and norms were not internally or externally consistent during the United States' federal response to the COVID-19 crisis. In contrast to SARS (Ding, 2009), COVID-19 prompted responses from the New York state government to communicate risks directly to the people. This contest was but one strategy that New York State used to share messages about preventative health behaviors such as mask wearing with its citizens. Throughout the uncertainty and communicative lapses during the COVID-19 pandemic, tactics used by individual communicators outside of institutions have filled the communication void (Colton et al., 2017). The pandemic prompted user-producers to fill communication gaps with their own science communication (Doan, 2021), as tactics often serve marginalized communities (Alexander & Edenfield, 2021; Colton et al., 2017). Because COVID-19 disproportionately affected African American, Latinx, and Asian/Pacific Islander people in New York City (Do & Frank, 2021), using tactics to communicate the importance of mask wearing took on new significance: "They [user-producers] are not, however, necessarily anti-institutional; they are willing to work within institutional strategies when it suits them, and to step outside those strategies when the occasion warrants" (Kimball, 2017, p. 3). The spring 2020 policy promoting mask wearing was a perfect time to unite tactics and strategies through cocreation as the State of New York did by sponsoring this PSA contest.

Broadening organizational focus in technical communication leads to greater opportunities and greater challenges for crowdsourcing procedural information. Federal and state governments enacted rules through strategies (Kimball, 2017) particularly during the first wave of the COVID-19 pandemic with stay-at-home orders and mask mandates for public safety (Walwema, 2021). Effective communication gave nonexperts narratives to persuade them to abide by safety guidelines. Ding (2009) portends problems with tactical technical communication from the United States government during early 2020 from escalating individuals as the COVID-19 pandemic continues: "[P]ossible access to alternative media for risk communication purposes only further complicates the matter by introducing issues of personal interests, ethical decision making, and civic action." (p. 344)

With the lack of official information from the United States government, citizens instead relied on the State of New York or on unofficial sources of information; however, the State of New York stumbled when sharing COVID-19 information (Doan, 2021). Because (mis)information spreads so quickly via social media, establishing ethos through tactics create public trust via narratives and metaphors (Holladay, 2017). Crowdsourcing crisis communication has potential for greater inclusion, coinciding with the social-justice turn in TPC.

Using Tactics During the Social-Justice Turn

The social-justice turn in technical communication (Jones et al., 2016) has focused on recognizing TPC's own contributions to structural inequalities (Jones & Williams, 2018) and place in righting existing injustices (Walton et al., 2019) through accountability (Colton et al., 2017). Social-justice research in TPC is emancipatory (Phelps, 2021) through beneficence: a drive to benefit and uplift all people, particularly the vulnerable (Colton et al., 2017). As a user-centered framework, TTC can provide tactics to further communicators' goals in their own communities, a "techne of marginality" (Shelton, 2019, p. 19). TTC thus "require[s] the use of supplementary ethical frameworks to identify whether a given tactic is ethical or unethical or somewhere in between" (Edenfield et al., 2019, p. 181), as strategies and tactics can cause ongoing harm (Sarat-St. Peter, 2017).

Tactics and strategies can be used for beneficence by sharing medical information about gender-transition care (Edenfield et al., 2019), mental health diagnoses (Holladay, 2017), and scientific descriptions in plain language (Pflugfelder, 2017). The COVID-19 pandemic heightened the stakes for TTC, as pandemic misinformation and its associated preventative health behaviors can quickly go viral on social media (Doan, 2021; Sleigh et al., 2021). Thus, the "ethics of tactics via vulnerability" (Colton et al., 2017, p. 65) provide tactics for technical communicators to communicate about preventative health behaviors that protect everyone, especially the most vulnerable, through instructional PSAs.

PSA Videos as Tactical Technical Communication

PSA videos translate technical information into procedures for audiences to follow with pithy slogans, such as, 1993's "Click it or ticket" (NC Department of Transportation Governor's Highway Safety Program, n.d.), about wearing seatbelts. PSAs are instructive texts that combine the literacies and skills of technical communication with the appeals of classical rhetorical theory (Bourelle, et al., 2015). As the rhetoric of proclamations, these public-facing genres communicate health risks in plain language to promote safe behaviors (Ding, 2009). These PSA videos show how tactics and strategies can work together within this genre.

PSA videos are tied to a specific kairos. In March and April 2020, New York City was the epicenter of COVID-19 cases in the United States. These PSA videos thread the needle of this temporality: temporary messaging with limited scope (Kimball, 2017) that draw connections with personalization and narration (Kimball, 2006). The State of New York structured this contest as a strategy to encourage cocreation from tactical technical communicators. Distributing the winning PSA videos drew on the strategy of free sharing, as TTC is "often provided out of pride of accomplishment and a willingness to share expertise and knowledge" (Kimball, 2017, p. 2). In short, creating these PSA videos about mask wearing was a social good, but it also gave the winning indie filmmakers exposure, complicating the relationship between users and the state. Professional production companies winning this contest obscures the relationship between tactics and strategies, as the opportunity to create health communication went to video-making experts.

New York's PSA Video Contest

As COVID-19 cases soared in New York during the spring of 2020, the government initiated a unique communications strategy: inviting New Yorkers to submit their own videos for consideration as the state's official PSA to promote mask wearing. Here, New York's strategies and users' tactics converged, operating within the guidelines and strategic framework provided. The Wear a Mask campaign show how strategies and tactics work together to promote both organizational and individual goals. This case study examines the five finalists, chosen from over 600 entries for the competition. Approximately 186,000 votes selected *We <3 NY* as the winning PSA and *You Can Still Smile* as the runner-up. The following section discusses how these PSAs were subsequently used by New York State as public-service announcements and analyzes the other three finalists.

Complicating our discussion of strategies and tactics are issues of power dynamics within the entries themselves. While a full examination of these dynamics is beyond our scope, we recognize that many of the PSA finalists were produced professionally rather than by individuals. Our discussion thus focuses on the interactions of organizational strategies and individual tactics, while articulating the submissions' complexities.

We <3 NY

The winning video, *We <3 NY*, centers on the idea of reciprocity. The beginning explains that New Yorkers have watched health-care professionals and essential workers do their jobs throughout the early part of the pandemic, and now it is time for everyday New Yorkers to do *their* job and wear a mask. By creating this contrast between the life-saving work of essential workers and wearing a mask, it is immediately clear that the work ordinary New Yorkers are being asked to do—wear a mask—is not a particularly difficult task.

The video claims that if New Yorkers want New York to open and remain open, everyone needs to do their part and wear a mask. This implies that wearing a mask to quell the pandemic is something New Yorkers must do together, requiring a shift from each person thinking about themselves to thinking about New York, the living entity. This sense of community action is conveyed in the simple but effective spoken lines: "I wear a mask to protect you; you wear a mask to protect me" and "When we show up in a mask, we're showing up for each other." The audio ends with "I love New York. We love New York," once again emphasizing the idea of community and the beneficent need to wear masks. The use of the simple, iconic phrase *I love New York* helps to make the PSA memorable, while the plain language throughout prompted public dissemination of the information (Pflugfelder, 2017). As the public spread of and buy-in to the PSA's message was a central goal of the competition, this PSA's effective use of plain language makes it an apt choice as the contest's winning video.

While this video's power derives from the words being spoken, the background images offer a compelling contrast. As the video begins, we see iconic New York landmarks: the Freedom Tower high above lower Manhattan, the Brooklyn Bridge, the Empire State Building, and Radio City Music Hall. After several seconds, the background images shift to showing everyday New Yorkers doing everyday things like exercising in their apartments, loading a mail truck, and standing in a grocery store. As dialogue calls on all New Yorkers to act together, images convey a similar message, portraying the work of both essential workers and those who are staying home. The ending shows these ordinary New Yorkers outside, integrating the PSA's call to action (to wear a mask) into their everyday activities.

This PSA exemplifies technical communication's major turn to public participation as it was solicited and distributed by New York State but created outside its sphere of influence. The result is a mediated relationship.

As user-producers, workers at Bunny Lake Films used tactics within the bounds of New York State's strategy, crafting messaging that suited their dual needs as the PSA's producers and as New Yorkers themselves to reach their New York audiences.

We Are Compassion. We Are NY

We Are Compassion. We Are NY begins by showing a child wearing a blue-and-white disposable mask that says, "We are," which is then removed to reveal a second mask that reads, "NY Tough." The boy walks over to an older man sitting on a park bench and hands him a mask, which he puts on right away. The boy's mask then reads, "Let's prove it." Finally, he removes this mask to reveal one final mask that reads, "Together."

Like many of the other finalists, this video highlights the difference a single individual can make. The contrast between a young child and an older man emphasizes that New Yorkers, despite their differences, are in this together. The child sees a man without a mask and walks over to give him one. In performing this small act, he has made a big difference in this man's life. Viewers are called upon to do the same, to work "together."

This video contains no dialogue. Rather, words are written on the masks. While the message is still being communicated with words, it's interesting to consider the materiality of the masks themselves. Here, masks serve as a symbol of the narrative that wearing a mask protects other people. While this ad may not have intended to make the mask itself rhetorical, mask wearing has become a symbolic act, as wearing a mask has become aligned with political affiliation and other social factors.

That Guy

That Guy shows an unmasked man in a train station. All the people waiting near him and moving past him wear masks. The website describes him as "the blissful unmasked foe who no one wants to be near and who the people of New York will avoid at all costs." This implies that other New Yorkers know better and are doing better. By describing and presenting him as a foe, it also implies that he is in the wrong—that not complying with the mask mandate is selfish.

By focusing on someone who is not doing the right thing, this video takes on the idea of defiance. While some people not wearing masks do so out of ignorance, some people do so to act against mandates or recommendations or to reclaim their own agency. While this PSA may seem to have a simple message, there are still tactics to consider here. Maybe "doing the right thing" is not enough to encourage people to comply; maybe fear of being "that guy" is a better tactic for reaching some New Yorkers.

Like Colton and colleagues (2017) claim, tactics call attention to the unsanctioned, highlighting the idea of resistance. "That guy" is resisting the mask mandate, while the PSA resists the strategies of the organization in power and their communications, the majority of which focus on mask wearing as the right, responsible thing to do. In contrast, this PSA tactically promotes mask wearing in a new and unusual way. Because the PSA is user-produced rather than sponsor-produced, it may enjoy more freedom in its messaging, separating itself from the other finalists.

Do the Right Thing

Do the Right Thing juxtaposes video of New Yorkers around the city, many of whom are holding signs thanking essential workers, with audio from Governor Cuomo's April 29, 2020, daily COVID briefing to emphasize the relationship between politicians and "the people." Governor Cuomo says, "Sometimes the people lead, and the politicians follow." This contrasts with the typical relationship where people follow rules set forth by politicians, while giving more responsibility and agency to the people.

By increasing people's agency, this PSA, like many of the other finalists, encourages the audience to "do the right thing." However, doing the right thing is not complying with rules and laws but rather using personal agency in responsible and productive ways. The people are, hypothetically, given the power to decide what they—and the "following" politicians will do next.

While Governor Cuomo would later resign, in November 2021, amid a sexual harassment scandal (Ferre-Sadurno & Goodman, 2021), during the early parts of the pandemic, he stood out as one of the most respected government voices. His speech draws upon this ethos, tactically articulated by the PSA creators and not the contest's government sponsor. This grants Governor Cuomo a more authentic ethos that grows out of New Yorker's portrayal of him and thus increases the PSA's credibility. Instead of the PSA producers working within the framework of strategies to better meet their own needs, they engage with those strategies as part of their overall tactic. This PSA subverts normal power relationships, using New Yorkers to simultaneously spread the state's official message and bulwark its governor and its platform.

You Can Still Smile

You Can Still Smile begins with individuals, presumably New Yorkers, naming the people for whom they wear a mask, such as "my children" and "my mama who's a health-care worker." Others name groups on the front lines such as "transit workers" and groups they deem to be most at risk, including "the immunocompromised" and "marginalized communities who don't have access to adequate healthcare." As Kimball (2006) notes in his foundational work on TTC, personalization and narration become tactics that communicators use to resist institutional authority. In some ways, then, this video enacts this idea, as the creators have deliberately chosen to highlight particular communities that they deem most at risk. Those featured in the video state other reasons they wear masks, highlighting their desire to go back to normal and to be able to interact with each other. There is an acceptance of responsibility—that their actions are the only way to bring normalcy back to New York.

The explicit acknowledgement of marginalized communities makes this ad a particularly interesting—and important—use of tactics. While tactics act as "an art of the weak" (Kimball, 2006, p. 71; de Certeau, 1984, p. 37), here, those tactics are used by other New Yorkers to give voice to these at-risk communities who have been most impacted by COVID-19 (Do & Frank, 2021). In doing so, they transform from passive to active agents of technical communication. Tactics here are not being acted out by the communities most at risk but rather on their behalf. In considering the user-producers here, this is one of only two of the video finalists not produced by a studio. The creators, then, truly act on behalf of themselves and other New Yorkers to spread the wear-a-mask message to promote public health for the greater good.

Implications

From our analysis, we draw three implications: promoting user agency within power dynamics, blending strategies and tactics to cocreate rhetorics of proclamations, and promoting beneficence through TTC.

Promoting User Agency Within Power Dynamics

These video submissions highlight the agency of everyday citizens by challenging typical power dynamics. The winning PSA was not produced by an individual, but rather by Bunny Lake Films. Thus, while the PSA gave voice to the health concerns of New Yorkers, the citizens' voices in this case were interpreted by the PSA's producers. Three of the five finalist videos were professionally produced, requiring us to reflect on the more robust resources of such companies over the average New Yorker and potentially complicating their motives to both represent New Yorkers and spread important health information, while gaining recognition.

This case study also highlights user-agency being enacted in unique ways and themes. These PSAs interpret the voices of New Yorkers through several different production companies (Bunny Lake Films; Skyline99 Studios; Plastic Tree Productions) and directly from the minds of fellow New Yorkers. They connect to their audiences by evoking a sense of responsibility and togetherness as seen in the *Do the Right Thing*, *We <3 New York*, and *You Can Still Smile* PSAs, by using contrast to highlight need as in the *We <3 New York* and *We Are Compassion* PSAs, and through a theme of defiance and an attempt to reclaim agency as seen in the *That Guy* PSA. The iconography of New York, like the Statue of Liberty and the Freedom Tower, enacts a rhetoric of locality, while the familiar disposable mask draws upon the now-ubiquitous image, communicating both its materiality and its range of contemporary meanings and associations. These PSAs portray a unique form of TTC, tied to the specific kairos of the early stages of the COVID-19 pandemic in New York. They use plain language to communicate time-sensitive health instructions, represent a willingness to share and circulate expertise, and use a rhetoric of proclamation (Ding, 2009) that focuses on imminent risk.

Blending Strategies and Tactics Through Cocreation

Most importantly, these PSAs blur the boundaries between strategies and tactics. They represent technical communication within the bounds of the institution in power, while also highlighting users' own priorities and concerns. These videos show how personalization and narration become tactics that resist institutional authority (Kimball, 2006), both directly in those videos created by individuals such as *You Can Still Smile*, and indirectly in those videos created by production studios such as *We <3 New York*. In both cases, users employed tactics to communicate within the boundaries set by the State of New York, leading to a kind of cocreation between various stakeholders with a shared kairos and intertwined agency.

Promoting Beneficence with Tactical Technical Communication

Promoting beneficence in TTC (Colton et al., 2017) remains important over both short-term and long-term crises. During COVID-19, these tactics were used to promote government policies that protect all people, such as stay-at-home orders or school closures (Doan, 2021). However, antigovernment groups used similar tactics to resist institutional authority and spread misinformation around mask wearing and vaccination against COVID-19 (Sleigh et al., 2021). To promote beneficence, technical communicators had to use narration and personalization to engage resistant audiences with facts about preventative health behaviors.

Conclusion

In PSA videos, strategies and tactics work together for beneficence. During times of crisis, top-down communication can successfully integrate the tactics of tactical technical communicators through cocreating the rhetoric of proclamations (Ding, 2009) to uplift marginalized perspectives in their own words. Although we expected to see these PSA videos use social justice as subtext, we argue that the kairos of Spring 2020 infused beneficence into these videos' texts themselves.

From these examples, technical communicators can learn that cocreation with institutional guardrails provides opportunities for reciprocity: people can highlight their own priorities for sponsoring institutions. Using the structure of a PSA video contest as a strategy, New York State highlighted user-producers' priorities and values, giving users agency in their own mask-wearing practices. Second, although PSA's such as "Click it or ticket" (NC Department of Transportation Governor's Highway Safety Program, n.d.) are ongoing, these masking PSAs highlight kairotic tactics and strategies. These PSA videos about mask wearing represent the social fabric of a time and place during a worldwide emergency; the winning videos used tactics to communicate this singular urgency. Finally, as these PSAs show, TTC embodies the social-justice turn by attending to power dynamics through narration. During the user-centered turn in technical communication, we must both represent users' words and credit, compensate, and recognize their participation. As COVID-19 continues, tactical technical communicators must pay closer attention to the techne of marginality (Shelton, 2019) that encourages tactical technical communicators to work towards beneficence.

References

Alexander, J. J., & Edenfield, A. C. (2021). Health and wellness as resistance: Tactical folk medicine. *Technical Communication Quarterly, 30*(3), 241–256. https://doi.org/10.1080/10572252.2021.1930181

Bell, I. (2020, May 19). *Do the right thing.* [Video]. YouTube. https://www.youtube.com/watch?v=gUsTfYHxSFs

Bellwoar, H. (2012). Everyday matters: Reception and use as productive design of health-related texts. *Technical Communication Quarterly, 21*(4), 325–345. https://doi.org/10.1080/10572252.2012.702533

Bougadellis, N., & Parker, E. (2020, May 19). *You can still smile* [Video]. YouTube. https://www.youtube.com/watch?v=7xawaYqckR0

Bourelle, A., Bourelle, T., & Jones, N. (2015). Multimodality in the technical communication classroom: Viewing classical rhetoric through a 21st century lens. *Technical Communication Quarterly, 24*(4), 306–327. https://doi.org/10.1080/10572252.2015.1078847

Bunny Lake Films. (2020, May 22). *We <3 York* [Video]. YouTube. https://www.youtube.com/watch?v=-l2SRDL26q4&t=2s

Colton, J. S., Holmes, S., & Walwema, J. (2017). From NoobGuides to #OpKKK: Ethics of anonymous' tactical technical communication. *Technical Communication Quarterly, 26*(1), 59–75. https://doi.org/10.1080/10572252.2016.1257743

Ding, H. (2009). Rhetorics of alternative media in an emerging epidemic: SARS, censorship, and extra-institutional risk communication. *Technical Communication Quarterly, 18*(4), 327–350. https://doi.org/10.1080/10572250903149548

Do, D. P., & Frank, R. (2021). Unequal burdens: Assessing the determinants of elevated COVID-19 case and death rates in New York City's racial/ethnic minority neighbourhoods. *Journal of Epidemiology and Community Health, 75*(4), 321–326. https://doi.org/10.1136/jech-2020-215280

Doan, S. (2021). Misrepresenting COVID-19: Lying with charts during the second golden age of data design. *Journal of Business and Technical Communication, 35*(1), 73–79. https://doi.org/10.1177/1050651920958392

Edenfield, A. C., Holmes, S., & Colton, J. S. (2019). Queering tactical technical communication: DIY HRT. *Technical Communication Quarterly, 28*(3), 177–191. https://doi.org/10.1080/10572252.2019.1607906

Ferre-Sadurno, L. & Goodman, J (2021, November 10). Cuomo resigns amid scandals, ending decade-long run in disgrace. *New York Times.* https://www.nytimes.com/2021/08/10/nyregion/andrew-cuomo-resigns.html

Henning, T., & Bemer, A. (2016). Reconsidering power and legitimacy in technical communication: A case for enlarging the definition of rechnical communicator. *Journal* of *Technical Writing and Communication, 46*(3), 311–341. https://doi.org/10.1177/0047281616639484

Holladay, D. (2017). Classified conversations: Psychiatry and tactical technical communication in online spaces. *Technical Communication Quarterly, 26*(1), 8–24. https://doi.org/10.1080/10572252.2016.1257744

Jones, N. N., Moore, K. R., & Walton, R. (2016). Disrupting the past to disrupt the future: An antenarrative of technical communication. *Technical Communication Quarterly, 25*(4), 211–229. https://doi.org/10.1080/10572252.2016.1224655

Jones, N. N., & Williams, M. F. (2018). Technologies of disenfranchisement: Literacy tests and black voters in the US from 1890 to 1965. *Technical Communication, 65*(4), 371–386.

Kimball, M. A. (2006). Cars, culture, and tactical technical communication. *Technical Communication Quarterly, 15*(1), 67–86. https://doi.org/10.1207/s15427625tcq1501_6

Kimball, M. A. (2017). Tactical technical communication. *Technical Communication Quarterly, 26*(1), 1–7. https://doi.org/10.1080/10572252.2017.1259428

NC Department of Transportation Governor's Highway Safety Program. (n.d.). *Click It or Ticket—History*. https://www.clickitorticket.org/Pages/history.aspx.

New York State. (n.d.). *Wear a mask New York ad contest: Winner announced*. Health. Ny.Gov https://coronavirus.health.ny.gov/wear-mask-new-york-ad-contest-winner-announced

Pflugfelder, E. H. (2017). Reddit's "explain like I'm five": Technical descriptions in the wild. *Technical Communication Quarterly, 26*(1), 25–41. https://doi.org/10.1080/10572252.2016.1257741

Phelps, J. L. (2021). The transformative paradigm: Equipping technical communication researchers for socially just work. *Technical Communication Quarterly, 30*(2), 204–215. https://doi.org/10.1080/10572252.2020.1803412

Plastic Tree Productions. (2020, May 19). *That guy*. [Video]. YouTube. https://www.youtube.com/watch?v=01xpKejzhhg

Reardon, D. C., Wright, D., & Malone, E. A. (2017). Quest for the happy ending to Mass Effect 3: The challenges of cocreation with consumers in a post-Certeauian Age. *Technical Communication Quarterly, 26*(1), 42–58. https://doi.org/10.1080/10572252.2016.1257742

Sarat-St. Peter, H. A. (2017). "Make a bomb in the kitchen of your mom": Jihadist tactical technical communication and the everyday practice of cooking. *Technical Communication Quarterly, 26*(1), 76–91. https://doi.org/10.1080/10572252.2016.1275862

Shelton, C. D. (2019). *On edge: A techne of marginality* [Unpublished doctoral dissertation]. East Carolina University.

Skykline99 Studios. (2020, May 19). *We are compassion. We are New York.* [Video]. YouTube. https://www.youtube.com/watch?v=J0hyE6MCC8c

Sleigh, J., Amann, J., Schneider, M., & Vayena, E. (2021). *COVID-19 on Twitter: An analysis of risk communication with visuals* [Preprint]. In Review. https://doi.org/10.21203/rs.3.rs-195119/v1

Walton, R. W., Moore, K. R., & Jones, N. N. (2019). *Technical communication after the social justice turn: Building coalitions for action* (First edition). Routledge.

Walwema, J. (2021). The WHO health alert: Communicating a global pandemic with WhatsApp. *Journal of Business and Technical Communication, 35*(1), 35–40. https://doi.org/10.1177/1050651920958507

Wilson, G., & Wolford, R. (2017). The technical communicator as (post-postmodern) discourse worker. *Journal of Business and Technical Communication, 31*(1), 3–29. https://doi.org/10.1177/1050651916667531

Chapter Five

Grassroots Activism and Tactical Communities

Examining the Poor People's Corporation in Mississippi in the 1960s and 1970s

Don Unger

This chapter focuses on the tactical technical communication (TTC) employed by a grassroots organization, the Poor People's Corporation (PPC). The PPC emerged from the Mississippi civil rights movement in 1965 as a series of cooperatives owned and operated by Black workers that existed until 1974. I begin by tracing the organization's history because it is impossible to understand their approach to TTC without knowing something about this history. Then, I use TTC as a framework to analyze a small set of PPC documents. In so doing, I characterize the PPC as a tactical community, expanding on an underresearched aspect of Kimball's (2006) work. Also, I point out some shortcomings in de Certeau's (1984) theories of strategies and tactics, which have undergirded important discussions of TTC. Overall, my goal in this chapter is to refine the concept of tactical communities by centering grassroots activism and activist organizations like the PPC and to make space for similar historical work in technical communication (TC).

The Poor People's Corporation: Its History and Structure

Before I describe how particular documents reflect tenets of TTC and how, taken together, they present the PPC as a tactical community, I provide a brief history of the organization. This history helps contextualize the documents by informing the reader about ongoing challenges that the organization faced and demonstrating how these documents reflect an important context for and approach to TC. To provide this history, I lean on the single published article that traces it, William Sturkey's (2012) "Crafts of Freedom: The Poor People's Corporation and Working-Class Activism for Black Power." Additionally, I supplement Sturkey's work with information from articles published by popular and alternative presses during the 1960s and 1970s, as well as archival documents by or about the PPC.

The Poor People's Corporation in Brief

In summer 1964, a group of Black women in Canton, Mississippi, got together to develop ideas for products that they could make and sell so that their families would gain financial independence from white employers who would retaliate against them for registering to vote. The women established the Madison County Sewing Firm, and they turned to Jackson-based civil rights activists in the Council of Federated Organizations (COFO) for help expanding their business (Sturkey, 2012). COFO was a coalition of several civil-rights-movement organizations operating at that time, including the NAACP, SNCC, and CORE (Congress of Racial Equality). The PPC grew out of this collaboration. Through COFO, the sewing cooperative connected with Jesse Morris, a Black activist and a SNCC field secretary from Florida, who had a degree in Economics from the University of California, Los Angeles. Morris served as the PPC's secretary and de facto director over the course of its history (SNCC Digital Gateway [SNCC], n.d. "Jesse Morris").

The PPC held its first membership meeting on Sunday, August 29, 1965, at Tougaloo College, with 300 people in attendance (Poor People's Corporation [PPC], 1965a). At the meeting, members presented business proposals, including start-up budgets and sample goods that their proposed co-op would produce. The first meeting distributed $4,895 to 9 co-ops, including the Madison County Sewing Firm (PPC, 1965a). Morris also argued for money to be spent on creating a co-op in Jackson, Mississippi that would sell wares from all the other co-ops. They named the Jackson venture the Liberty House Co-op (Sturkey, 2012).

The money divvied out at this and subsequent annual meetings came in the form of loans. Bylaws stated that co-ops had to repay these loans, interest free, within specified time periods. Additionally, every member paid 25 cents as yearly dues. Loan repayments and dues were then added to the "revolving fund" and used to fund future proposals, some of which helped bolster existing co-ops, and some went toward creating new ones. Additionally, Morris traveled around the country to secure donations from sympathetic organizations and individuals. While groups who placed orders with the PPC paid for the products with grant funds secured from private organizations and state and federal agencies, Morris opposed the idea of the PPC taking funding directly from the government (Sturkey, 2012).

Early on, many co-ops secured their own contracts with retailers, but as the PPC grew, it sought to make their goods centrally available. Managed by Doris A. Derby, a SNCC field secretary from 1963 through 1972, the Liberty House Co-op served as the glue that connected the co-ops together, and it comprised a distribution center, the Education and Training Center, and a storefront (SNCC, n.d., "Doris Derby"). Located at 313 Pascagoula St. in Jackson, Mississippi, the distribution center ordered materials in bulk for co-ops across the state and handled orders and marketing for PPC products. The Education and Training Center coordinated workforce training for co-op members, hosting artisans from across the country who helped lead this training. The storefront, dubbed the Liberty House Outlet and located at 614 N. Farish Street in Jackson, sold PPC goods to local customers. Additionally, the Liberty House Co-op marketed PPC goods at trade shows, and in the late 1960s, Liberty House Co-op Outlets opened in New York, Detroit, Chicago, and Berkeley. These storefronts were run by Friends of SNCC members, such as Ellen Maslow and Abbie Hoffman (Sturkey, 2012). In 1966, 27 retailers from San Francisco to Grinnell, from Indianapolis to Pompano Beach, and beyond were also selling PPC goods (Sturkey, 2012).

Despite the fact that the PPC continued to grow throughout the late 1960s, all but two of the initial co-ops established at the first membership meeting in 1965 dissolved by 1967 (Sturkey, 2012). PPC members depended on their co-ops turning a profit each quarter, and when a co-op couldn't, it fell apart, or given strong local leadership and luck, it found temporary tactics to buoy members' incomes. Where Liberty House once served as the glue that kept the PPC co-ops together, it became a hindrance to some co-ops' successes because it tied the work across the co-ops to one another based on how the PPC procured raw materials: if Liberty House fell behind in ordering and delivering these materials, then it meant that orders across the co-ops fell behind, and with that, sales and income fell behind (Sturkey, 2012). In 1970, the Liberty House filed for bankruptcy despite the fact that it reported gross sales of $119,492.43 (approx. $914,080 in 2022); a year later, it was $24,985.77 in the red (Sturkey, 2012). Finally, as Sturkey (2012) notes, "by 1973, many of the co-ops had folded, and in 1974, the organization was forced to liquefy all its holdings" (p. 58).

To understand the PPC's collapse, one needs to understand the sheer enormity of its project as well as the context in which the organization operated. In considering this context, one cannot overemphasize the role that the state of Mississippi played in sabotaging its Black residents and Black political, cultural, and economic organizations operating within the state, including the PPC.

In 1956, two years after the *Brown v. Board of Education of Topeka, Kansas* decision, the Mississippi State Legislature established the Mississippi Sovereignty Commission with the objective of "do[ing] and perform[ing] any and all acts and things deemed necessary and proper to protect the sovereignty of the state of Mississippi, and her sister states" from "encroachment thereon by the Federal Government or any branch, department or agency thereof; to resist the usurpation of the rights and powers reserved to this state and our sister states by the Federal Government or any branch, department or agency thereof" (Rowe-Sims, 2002, para. 3). In practice, this meant surveilling, intimidating, and harassing civil-rights-movement activists.

Archival documents from the commission show that the PPC was a significant target. Some documents address the commission's surveillance and intimidation tactics on individuals. Such documents profile PPC leaders and track their movements across the country, discussing their relationships with businesses and people inside and outside Mississippi (Mississippi Sovereignty Commission Archive [MSCA], 1966a, 1966b). Others list the home addresses and telephone numbers of meeting participants (MSCA, 1966c). Finally, some of the commission's documents include internal memos attacking organizations that were PPC customers, such as the Child Development Group of Mississippi (MSCA, 1966d).

While the commission's attacks on the PPC demonstrate some of the challenges that the organization faced, these documents do not reveal the violence unleashed on Black Mississippians, particularly on those associated with the civil rights movement. However, the PPC's documents do. Many of the PPC's documents balance writing conventions still common to businesses and workplaces today with personal narratives, interviews, and testimonials that speak to the conditions that Black Mississippians faced in their daily lives. For example, the PPC distributed a number of documents to members and potential members at their meetings. In 1965, these documents included meeting minutes and agendas alongside documents like "Up from Mississippi—Mrs. Glover: 'I'm Really in the Movement Now' " (1965) about the experiences of a Black Batesville, Mississippi school teacher as well as a "Partial List of Racial Murders in the South in the Last 2 Years" (1965). These documents detail abuse at the hands of white employers and neighbors as well as the police. Considering the relationships among these documents points toward the PPC's lasting impact.

In addressing this impact, Sturkey (2012) points out that the organization's history challenges dominant narratives about the Black Power Movement in the United States. He argues that "most southern grassroots economic programs in the mid-1960s South have been . . . excluded from the Black Power narrative" as this narrative focuses on Black men operating in northern urban centers (p. 27). Instead, the Black Power exemplified by the PPC reflects "organic, female-based community organizing principles that had existed among working-class African Americans for generations" (p. 27). In the end, Sturkey (2012) believes that, had things gone differently for Liberty House, the PPC could have "resulted in a truly remarkable American Black Power story, one that offers a fundamentally different narrative of African American late-1960s economic and political organizing" (p. 60).

I contend that the PPC's lasting impact as a grassroots organization based on Black radical traditions points toward new avenues for research that could fundamentally change TC scholarship, avenues that explode the traditional starting point for the field: embedded in whiteness, in corporations, and in the ways that employees use writing to inculcate users to capitalist, consumerist paradigms or the ways that employees and users develop tactics to transgress these strategies. By learning about the PPC's history, we see a different starting point for TC. In examining how the organization's documents interweave technical and business content with personal narrative and content related to the civil rights movement, we see the field from a vantage point that centers Blackness and anticapitalist beliefs and practices.

The PPC and TTC

To clarify this perspective, I turn to scholarship on TTC in order to examine the tactics the PPC used in a small set of documents. In this section, I discuss three documents that the PPC circulated to its members in 1965 as well as a brochure proposing the Mississippi Brick Co-op. In my analysis, I illustrate how the documents reflect practices that Kimball (2017, 2006) draws from in order to develop his theory of TTC, specifically *la perruque*, *bricolage*, and radical sharing. Finally, I conclude the section by discussing how, taken together, these documents point toward the PPC as a tactical community (Kimball, 2006).

In describing the ways that users develop TC that transgresses strategies imposed on them by corporations, organizations, and institutions, Kimball (2017, 2006) identifies three tactics. He draws two, *la perruque* and bricolage, from de Certeau (1984), and he offers a third tactic derived from his own research, radical sharing. *La perruque* involves "appropriating time or surplus material at work to personal uses" (Kimball, 2006, p. 72). By way of examples, de Certeau (1984) offers one of a cabinetmaker borrowing a lathe from his employer in order to use it at home to make furniture and another of a secretary writing a love letter on company time (p. 25). At its base, this tactic diverts resources away from the boss's profit margin in order to accomplish something meaningful for the worker. Bricolage is often translated as "making-do," as it refers to "making and doing what you can with what you have" (Kimball, 2006, p. 71). Put another way, the term refers to the often hidden, individualized tactics that people develop in order to work with or around the strategies imposed by institutions, organizations, the state, and their representatives. Finally, Kimball (2017) defines radical sharing "as our newfound individual capability of sharing our tactics with people the world over at great speed and with great effect" (p. 4). He links this newfound ability to the internet and digital tools that allow individuals to share their tactics with global audiences and differentiates radical sharing from something like citizen journalism: it is "not about what happened, but about making things happen" (Kimball, 2017, p. 4). By reading the PPC documents for evidence of these tactics, I expand our understanding of the tactics, and I attempt to work through some contradictions in de Certeau's (1984) theories.

The PPC Member Documents

The small set of documents that I look at were distributed to PPC members in 1965. They include a document titled "The Poor Peoples Corp.," which details the organization's formation and structure (PPC, 1965b). Next, this set of documents includes a personal narrative retyped from a 1965 *Village Voice* article about Mrs. Thelma Glover, a 46-year-old Black school teacher from Batesville, Mississippi ("Up from Mississippi," 1965). Finally, I look at a document titled "Partial List of Racial Murders in the South in the Last 2 Years" (1965).

Traditionally, scholars often regard *la perruque* as a tactic for getting one over on bosses or the businesses that these bosses represent (e.g., Kimball, 2006). However, this tactic is sometimes assumed to be implemented in individualistic ways. "The Poor People's Corporation" document aims to get one over on bosses (or former employers), but it takes a slightly different approach. It aims to take the workers away from their employers entirely. It begins with a page and half description of how the PPC works, using a sewing co-op as an example. This description concludes with an argument on why people should join the PPC, which reads:

> This is just one example of what can be done to provide jobs for local people who cannot make enough on farms or as domestics, or for people who walk away from their jobs because they are on strike or lose jobs because of being active in the Movement. There are countless ways groups can get together and request assistance from the Poor Peoples Corporation. The job for all of us is to get together and decide if we all want to join the Poor Peoples Corporation and what type of aid we want. (p. 2)

The document's remaining four and a half pages follow a Q+A format, explaining how to create a co-op and join the PPC, how the loan process works, and the organization's structure. As this document shows, the PPC's tactics are localized but not individualistic because they encourage community members to work together to form co-ops. In fact, PPC materials frequently push back against individualism and emphasize collective well-being and action.

The next document, "Up from Mississippi—Mrs. Glover: 'I'm Really in the Movement Now'" (1965), tells how Mrs. Glover is deeply connected to her family, her community, her people, and the movement. The piece begins by discussing a trip she took to New York City to meet with movement organizers there. She had just been released from a makeshift prison on the Jackson, Mississippi, fairgrounds where police incarcerated 745 people who had attempted to march to the state capital on June 14, 1965. In the story, Mrs. Glover employs *la perruque* in relation to the situation, stating, "You know since I went to jail I'm really in the movement now. But I'm not scared of the sheriffs and the beatings—it's all the people I owe money to—almost $2000. But I guess they'll see to it that I don't lose my teaching job" (p. 4). Mrs. Glover's tactics for dealing with her arrest relate to getting one over on her employer, and the tactic is intertwined with her relationships to others. Furthermore, in describing her experiences in the makeshift prison, she recounts a story about how those arrested dealt with harassment from their jailers: "One morning, the police were beating nosily with their night-sticks—over the loud-speaker. It had become a common form of harassment, but this time a group of girls began to do African dances to the beating sound" (p. 3). The incident illustrates bricolage in the sense that the women refused to succumb to the police officers' strategy of terror and harassment. While scholars often interpret bricolage to mean the individual tactics used to work with or around institutional strategies, Mrs. Glover points toward how the tactic can be a social rather than an individual response.

Taken together, the documents show various ways that Black Mississippians might get one over on their bosses or make do in spite of the horrific violence and continual harassment that they faced from the white supremacist police officers and others. Additionally, the documents point toward radical sharing as a way to help oneself *and* others do something—to create a co-op, to join the PPC, to advance the movement. They also advance this communal sensibility by sharing information that is meant to evoke an emotional response and not simply to inform people as in citizen journalism. The final document in this small set helps make this clear.

The "Partial List of Racial Murders in the South in the Last 2 Years" (1965) includes 24 cases—some involving multiple murders, in Mississippi, Alabama, Louisiana, and Georgia. The list documents highly publicized murders of movement leaders, such as Medgar Evers' murder in June 1963 and Michael Schwerner, James Cheney, and Andrew Goodman's murders during Freedom Summer in June 1964, but it also includes murders where the victims' names were not identified. In each case, the document includes the date, the victim, a description of the victim, the location of the crime, and a description of the crime. At the end of each listing, the anonymous authors include information about whether or not anyone had been arrested and/or convicted for the crime. In 16 of the 24 listings, arrests were made. In 22 of the 24 cases, "NO CONVICTION" was made (emphasis in original). The remaining two cases from March 1965 were marked as "PENDING" (emphasis in original). While the document might well provide readers with information in a journalistic sense, the visual emphasis on arrests and convictions evokes a visceral response. Also, listing movement leaders alongside anonymous victims illustrates Black peoples' shared experience of white supremacist violence.

The radicalness of the document relies on its connection to the others in the set, which not only encourage people to join the movement, but they also encourage people to join the PPC. The introduction to the Q+A in the first document makes this clear: "The poor are going to be given an opportunity to create jobs for themselves and develop industries of their own—industries that will be highly decentralized, labor intensive, utilizing not return on capital as a guiding incentive, but return to labor; industries planned, controlled, and operated solely by the people who work in them" (p. 2). This communal sense of radical sharing permeates the many documents circulated by the PPC and individual co-ops.

The Mississippi Brick Co-op Brochure

The Mississippi Brick Co-op, alternately dubbed the Freedom Now Brick Company, was an initiative directed by Frank Smith. Smith joined the Mississippi civil rights movement in 1962 as a SNCC Field Secretary, prior to finishing his education at Morehouse College (SNCC, n.d., "Frank Smith"). In 1965, he also helped found the Mississippi Freedom Labor Union (MFLU) (SNCC, n.d., "Mississippi Freedom Labor Union funded") in Shaw, Mississippi. Throughout 1965 and 1966, the MFLU organized farm workers throughout the Delta. At its peak, membership included some 1,350 workers across the region (SNCC, n.d., "Mississippi Freedom Labor Union funded"). Throughout its history, members carried out strikes in various towns aimed at winning the minimum hourly wage for farm workers, $1.25/hour. MFLU actions culminated in 1966 with Strike City when farm workers and their families took over the US Air Force Base in Greenville, Mississippi. The protestors occupied the space for months, living in tents as they began to build homes on the property (Mississippi Civil Rights Project, n.d.). Ultimately, the occupation ended when air force police forcibly removed the protestors (SNCC, n.d., "February 1966").

While Smith supported the MFLU organizing and action, he also worked on the brick co-op. The co-op was incorporated in August 1965 and began holding planning meetings in September 1965 ("Brick Factory Meeting," 1965). Plans for the co-op included a workforce training program, and Smith reached out to companies and individuals around the country to share their expertise and to help fund it ("Brick Factory Meeting," 1965).

This particular brochure, however, aims at garnering support from Black Mississippians (Smith, n.d.). It serves as a visual proposal, one meant to speak to families like those who participated in Strike City. With that context in mind, we see how the brochure mingles *la perruque* and bricolage by depicting bricks as integral to getting one over on the powers that be and to making do. As Strike City illustrated, Black workers relied on white farm owners for both housing and income. Once the workers in the MFLU went on strike, they no longer had homes, much less jobs. A sensitivity to this dynamic permeates the brochure beginning with the cover, which reads, "Bricks for Freedom: A Co-op Proposal" at the top and " 'Not to make money, but to build foundations' " at the bottom (Smith, n.d.). The cover also shows an illustration of a Black man laying bricks. The body of the brochure is essentially a sequential narrative that juxtaposes captions with illustrations. The first page begins, "Some of the things people need are . . ." (Smith, n.d.). This introduction is followed up with illustrations and attendant descriptions of houses, hospitals, farms, and gardens. Much of the rest of the brochure extends the argument that a brick factory can meet many of Black Mississippians' needs by showing how the co-op would support the creation of both buildings and opportunities. In addition to hospitals and houses, the body of the brochure adds schools, factories, and other co-ops to the list of institutions that would benefit (Smith, n.d.). The brochure exemplifies how the PPC and its co-ops extend radical sharing beyond one context to include potentially everyone in Black communities. This community-oriented sense of radical sharing is not only a tactic to achieve a specific goal, such as creating jobs; it is the goal itself, creating freedom by lessening or completely removing Black Mississippians' dependence on racist white people.

The PPC's complex and community-oriented approach to *la perruque,* bricolage, and radical sharing prompt me, on the one hand, to reconsider the ways in which tactics undergird TTC. On the other hand, they prompt me to reconsider some basic tenets of de Certeau's (1984) theories. Such reconsideration might begin by examining Kimball's (2006) use of the term *tactical communities.*

The PPC as a Tactical Community

One challenge I face in using de Certeau's (1984) concepts to analyze PPC documents results from the philosopher's focus. As he notes in the introduction to *The Practice of Everyday Life*, he is interested in " 'popular culture' or marginal groups" and not " 'counter-culture' groups," which he writes off as "often privileged, and already partly absorbed into folklore" (de Certeau, 1984, p. xii). The PPC itself might fall in between the latter two of these three groups. Certainly, PPC members were from "marginal groups," but they were also countercultural in that they were systematically excluded from "popular culture" because they were Black, *and* their work sought to counter white supremacist culture. Colton, Holmes, and Walwema (2017) identify an issue that overlaps with my concerns here. The authors argue that the "ethic of tenacity" that undergirds de Certeau's (1984) theory always already positions the "weak" as ethical. For their part, the authors seek another basis from which to assess acts of TTC as ethical or unethical. I am not concerned with whether or not the "weak" are always ethical. I am, however, concerned with how de Certeau (1984) positions the "weak" individual as always in some way in opposition to the group, the community, the organization. The PPC practiced radical democracy, and co-op owners were also the workers. While the PPC had leaders and centralized resources—the policies around which contributed to the organization's downfall, its dissolution was not preordained because the PPC was an organization. As I have demonstrated, the PPC and Black Mississippians in general were subject to extreme oppression both from the state and from white business owners and neighbors. De Certeau's (1984) general notion that institutions create buffers to change loses much of its explanatory power in the PPC's case. In this case, the organization was an epicenter for change. Rather than parsing the terms *organization* or *institutions* to further clarify this point, I turn to Kimball's (2006) work on tactical communities because it helps move the discussion past the bifurcation of institutions and strategies on one hand and individuals and tactics on the other.

Kimball (2006) works through the seeming impasse of institutions/strategies and individuals/tactics by advancing the concept of tactical communities. Drawing from de Certeau's (1984) theories discussed previously and melding them with Johnson's (1998) work on user-centered technology design, Kimball (2006) describes how tactical communities emerge when groups of users develop "extra-institutional systems" based on their practices scavenging, altering, or combining disparate and discarded technologies (p. 82). Such systems produce and circulate user documentation that develops similarly. Members of these tactical communities "with no place per se" have moved from the positions of user-as-consumer and user-as-producer to user-as-citizen (Kimball, 2006, p. 82). In Johnson's (1998) framework, *user-as-citizen* refers to those folks whom professional designers should consult in developing and documenting products (p. 61). But in considering this framework in the context of the PPC, Johnson's (1998) notion of the user-as-citizen relies on insiders with a certain degree of institutional power valuing others who are outside the institution, the powerless, and making the dominant institution better. Clearly, this perspective would be a pipe dream in terms of addressing the lived experience of Black Mississippians in the 1960s. However, the discussion and concept illustrate the focus of traditional TC. While Johnson's (1998) work broke new ground when it was published, it also reified that idea that TC centers around institutions—even when technical communicators seek to make those institutions more open, responsive, or democratic. In rethinking tactical communities from the perspective of the PPC, that is to say, as a rival to dominant institutions, we can retain the concept of citizen, even as we jettison the term. The users-as-citizens can function in tandem with or as systems designers because of their experience, both individually and as a group (see Kimball's [2006] example of the Locost builder community). Citizens are recast as designers. They are intimately familiar with the systems in which they participate, including the inequities and failures.

Considering the PPC as a tactical community that positions the user-as-designer at the center reveals some lessons about what tactical communities (can) do. To conclude this section, I lay out some of these lessons. Tactical communities can reflect the TC work of folks who are excluded from corporations and dominant institutions but who have come together to create "extra-institutional systems" or rival institutions in order to meet their everyday needs. In terms of the PPC's approach, this means:

1. Tactical communities do not necessarily seek redress from those in power. Tactical communities can create institutions to rival dominant institutions. While many organizations who partnered with the PPC sought redress from white employers, the state, and the federal government, such as the Child Development Group of Mississippi or the Mississippi Freedom Labor Union, the PPC shunned such involvement.

2. In their writing and documentation, tactical communities blend personal narratives into traditional TC genres, juxtapose various genres alongside one another, or create new genres that reflect the community's beliefs, values, and experiences and help the community do something. At times, PPC documents appear to be similar to traditional TC genres used by corporations and dominant institutions, such as

proposals, memos, meetings minutes, and so on. When you dig into these documents, however, they are very different. The PPC's writing conventions meld traditional forms of TC with conventions and genres from the civil rights movement and other Black radical traditions, including the cooperative movement. For example, stories and personal experiences figure as prominently as technical descriptions.

3. Tactical communities use bricolage, *la perruque*, and radical sharing as community-wide tactics to deal with, resist, and challenge systemic oppression. They are not individualized tactics or hacks. They are community standards. Still, so long as the oppressive conditions that gave rise to these tactical communities exist, these standards do not develop into strategies, that is to say, they do not become the new oppressive conditions that the tactical community must grapple with. That is not to imply that tactical communities are infallible or that these standards do not need to be redressed when they isolate different people or groups of people in the tactical community. However, they are not equivalent to the strategies advanced by dominant institutions.

Conclusion

TTC and tactical communities have existed as long as what we would identify as TC has existed. What has changed is the willingness of those in the field to include TTC in our discussions. Scholarship on TTC has helped widen the space in the field where scholars might focus more explicitly on TC that has nothing whatsoever to do with what is all too often framed as making corporations more responsive to users' needs, or put another way, creating kinder, gentler corporations.

On the other hand, TTC relies heavily on de Certeau's (1984) theories of strategies and tactics. There is a problem with using de Certeau's work to discuss the activism of Black Mississippians during the civil rights movement instead of using frameworks from Black radical traditions. For example, relying on de Certeau (1984) necessarily restricts the ways in which I understand and document the PPC's work. It shapes what I look for, what I discuss, and what I find meaningful enough to convey to others. There are few conversations in TC where Black scholars are well represented and where the field draws from Black radical scholars and traditions for critical frameworks. There are certainly signs that the conversation is beginning to emerge, but it is hardly at the center of the discipline. Presently, joining a scholarly conversation in TC, even one that focuses on work by Black people, means making it accessible, meaningful, and useful to white frameworks in very specific ways. Therefore, as a white TC scholar carrying out this work, it is imperative to go beyond the predominant scholarly conversations in TC and TTC and locate myself in relation to this work. In this process of locating myself, I draw on Black feminist and womanist traditions. Layli Phillips and Barbara McCaskill's (1995) argue, "Perhaps the central organizing principle of womanism (if it can that there is one) is the absolute necessity of speaking from and one's own experiential location and not to or about someone else's (p. 1010)."

I understand that the PPC's history is not my story. Yet, as Sheryl Conrad Cozart and Jenny Gordon (2006), drawing on the work of Tamara Beauboeuf-Lafontant (2002), remind us: teachers and scholars need "to decide if the purpose of education is system maintenance or social transformation" (Cozart and Gordon, 2006, p. 13). Being a professor at the University of Mississippi and as a queer person working toward social transformation demands that I critically engage with both my university's and the state of Mississippi's racist history. It requires that I learn this history and that I address it in my research and teaching in order to contribute to contemporary efforts that fight back against it. It means learning this history and the ways that it continues to shape state and university strategies as well as how this history informs the tactics that people use to resist those strategies, particularly Black workers and students. I do not approach the PPC's history as my story but through a sense of solidarity in the ongoing struggle for civil rights in Mississippi. Future work on tactical communities should continue to examine TC developed by grassroots organizations based on Black radical traditions, and TC scholars should seek to play a supporting role in contemporary struggles. To do any less would simply contribute to "system maintenance."

References

Beauboeuf-Lafontant. T. (2002). A womanist experience of caring: Understanding the pedagogy of exemplary Black women teachers. *The Urban Review, 34*, 71–86. https://doi.org/10.1023/A:1014497228517

Brick factory meeting, September 10, 1965, Washington, DC (1965). Lucille Montgomery Papers, 1963–1967, Wisconsin Historical Society, Madison, Wisconsin. https://content.wisconsinhistory.org/digital/collection/p15932coll2/id/34309.

Colton, J. S., Holmes, S., & Walwema, J. (2017). From NoobGuides to #OpKKK: Ethics of Anonymous' tactical technical communication. *Technical Communication Quarterly, 26*(1), 59–75. https://doi.org/10.1080/10572252.2016.1257743

Cozart, S. C., & Gordon, J. (2006). Using womanist caring as a framework to teach social foundations. *The High School Journal, 90*(1), 9–15. http://www.jstor.org/stable/40364215

de Certeau, M. (1984). *The practice of everyday life.* University of California Press.

Johnson, R. R. (1998). *User-centered technology: A rhetorical theory for computers and other mundane artifacts.* State University of New York Press.

Kimball, M. A. (2017). Guest editor's introduction: Tactical technical communication. *Technical Communication Quarterly, 26*(1), 1–7. https://doi.org/10.1080/10572252.2017.1259428

Kimball, M. A. (2006). Cars, culture, and tactical technical communication. *Technical Communication Quarterly, 15*(1), 67–86. https://doi.org/10.1207/s15427625tcq1501_6

Mississippi Civil Rights Project. (n.d.) Strike City. https://mscivilrightsproject.org/washington/place-washington/strike-city/

Mississippi Sovereignty Commission Archive. (1966a). Jessie Turner Morris, aka, Jesse Morris. *Mississippi Department of Archives and History.* https://da.mdah.ms.gov/sovcom/images/png/cd02/009031.png and https://da.mdah.ms.gov/sovcom/images/png/cd02/009032.png

Mississippi Sovereignty Commission Archive. (1966b, December 1). Poor People's Corporation—Gerity Broadcasting Company. *Mississippi Department of Archives and History.* https://da.mdah.ms.gov/sovcom/images/png/cd10/076636.png

Mississippi Sovereignty Commission Archive. (1966c). Poor People's Corporation meeting in New York City on April 4, 1966. *Mississippi Department of Archives and History.* https://da.mdah.ms.gov/sovcom/images/png/cd07/049596.png

Mississippi Sovereignty Commission Archive. (1966d, December) Proposed statement to be issued by Governor Paul Johnson in connection with grant of $8,000,000 to Child Development Group of Mississippi. *Mississippi Department of Archives and History.* https://da.mdah.ms.gov/sovcom/images/png/cd10/079136.png and https://da.mdah.ms.gov/sovcom/images/png/cd10/079137.png

Partial list of racial murders in the South in the last two years. (1965). Poor People's Corporation Records, 1960–1967, Wisconsin Historical Society, Madison, Wisconsin. https://content.wisconsinhistory.org/digital/collection/p15932coll2/id/2987

Phillips, L., & McCaskill, B. (1995). Who's schooling who? Black women and the bringing of the everyday into academe, or why we started "The Womanist." *Signs, 20*(4), 1007–1018.

Poor People's Corporation. (1965a). Minutes of the first membership meeting of the Poor People's Corp. [Brochure]. Lucille Montgomery Papers, 1963–1967, Wisconsin Historical Society, Madison, Wisconsin. https://content.wisconsinhistory.org/digital/collection/p15932coll2/id/34543

Poor People's Corporation. (1965b). The Poor People's Corp. Poor People's Corporation Records, 1960–1967, Wisconsin Historical Society, Madison, Wisconsin. https://content.wisconsinhistory.org/digital/collection/p15932coll2/id/2983

Rowe-Sims, S. (2002, Sept.). The Mississippi State Sovereignty Commission: An agency history. *Mississippi History Now: An Online Publication of the Mississippi Historical Society.* https://mshistorynow.mdah.ms.gov/issue/mississippi-sovereignty-commission-an-agency-history

Smith, Frank. (n.d.) Bricks for freedom: A co-op proposal. Wisconsin Historical Society. https://content.wisconsinhistory.org/digital/collection/p15932coll2/id/34402

SNCC Digital Gateway. (n.d.) Doris Derby. https://snccdigital.org/people/doris-derby/

SNCC Digital Gateway. (n.d.) February 1966: Occupation of Greenville Air Force Base. https://snccdigital.org/events/occupation-of-greenville-air-force-base/

SNCC Digital Gateway. (n.d.) Frank Smith. https://snccdigital.org/people/frank-smith/

SNCC Digital Gateway. (n.d.). Jesse Morris. https://snccdigital.org/people/jesse-morris/

SNCC Digital Gateway. (n.d.) Mississippi Freedom Labor Union founded. https://snccdigital.org/events/mississippi-freedom-labor-union-founded/

Sturkey, W. (2012). Crafts of freedom: The Poor People's Corporation and working-class activism for Black power. *The Journal of Mississippi History, 74*(1), 25–60.

Up from Mississippi—Mrs. Glover: "I'm really in the Movement now." (1965). Poor People's Corporation Records, 1960–1967, Wisconsin Historical Society, Madison, Wisconsin. https://content.wisconsinhistory.org/digital/collection/p15932coll2/id/2983

Chapter Six

The Streets Have Eyes

"Copwatching" as Tactical Citizen Engagement with Police Policy

Michael Knievel

High-profile incidents such as the death of George Floyd in early summer 2020 concentrate the public's attention on police behavior, especially with regard to the use of force. Floyd's tragic death, sadly, falls within a lengthy and familiar sequence of controversial police-involved deaths involving deadly force used against African American men in recent years (e.g., Eric Garner, Michael Brown, Tamir Rice, Laquan MacDonald, Walter Scott Jr., Philando Castille, and numerous others), creating space for tactical citizen interventions into the execution of state/institutional power within the context of policing. Frustrated by institutional torpor and unsatisfied with other measures to increase police accountability regarding the use of force (see, for instance, *The President's Task Force on 21st Century Policing*, 2015), some citizens have sought alternative paths to reform at a microlevel by engaging police actions through "copwatching," the citizen-sponsored monitoring of police behavior.

Copwatching is sometimes taken up by citizen collectives or individuals interested in challenging the status quo through use of what Michel de Certeau (1984) calls "tactics," which citizens use to work around or to resist broader institutional "strategies" designed to organize society and promote certain outcomes. Tactics enable citizens to leverage opportunities to engender alternative outcomes at local, oftentimes personal levels, and the conceptual terrain surrounding them has increasingly inspired scholars of technical communication who are interested in exploring spaces where technical communication flourishes outside of organizations and institutions (e.g., Kimball, 2006; Ding, 2009; Holliday, 2017; Pflugfelder, 2017; Sarat-St. Peter, 2017; Van Ittersum, 2014). Related inquiry has frequently centered on ways in which these tactical interventions both contribute to and reshape contours of technical information, integrating and aggregating citizen knowledge in pursuit of citizen-determined goals. Scholars engaged in this conversation often describe citizens who identify gaps in institutionalized information and strategies ostensibly designed to serve the public good (sometimes described as the user or consumer good) but that are insufficiently responsive to individual or local needs, spurring citizens to innovate and otherwise constellate resources (e.g., Bellwoar, 2012; Finer, 2016; Petersen, 2018).

In the case of copwatching, citizens engage tactically with the invention and use of technical information in policing. Beyond the organizational policies that frame police action, the streets themselves serve as critical sites for many of policing's communication practices, as officers interact in real time with citizens and then assimilate and synthesize information to make decisions regarding the use of force. In this chapter, I describe the tactical technical communication work that copwatchers engage as they seek to complicate contemporary policing and refashion it at the microlevel. Through description and analysis of tactics, I argue that copwatchers undertake efforts to sidestep and/or manipulate extant institutional practices in order to evolve a more equitable and just form of policing, working outside of conventional structures of participation to assert citizen influence onto policing practices. To develop this argument, I draw from contemporary research and media accounts of copwatching, as well as from interviews with and observations of copwatchers.

History of Policing: Establishing Strategy, Occasioning Tactics

According to de Certeau (1984), to counter institutional control, citizens engage in "tactics," which emerge out of small, unregulated spaces within a broader, regulated terrain, introducing into that terrain "a way of turning it to their advantage that obeys other rules and constitutes something like a second level interwoven into the first" (p. 30). Kimball (2017) cites related, productive tension that these interwoven levels might introduce when he notes how "competition" between "nimble citizens" and their tactical activities, and "slow-changing institutions," can be impactful as such competition "gives individuals a chance to nudge institutional pathways toward human needs, while giving institutions the power to regulate changes that might not be good for everyone" (p. 5). Other technical communication scholars have isolated complexes of activity that embody, in different ways, this competitive tension across institutional and tactical civilian-driven texts. For instance, Bellwoar (2012) describes ways in which tactical citizens "laminate" combinations of institutional texts with personal texts and experiential knowledge to navigate the healthcare system and the challenges of a complex pregnancy (p. 331). Other scholars, like Pflugfelder (2017), highlight ways in which tactical communication efforts add shape and contour to existing technologies and technical documentation (pp. 29–30). Frequently, tactical communicators come together in spaces such as user forums in order to pool their knowledge and experience, allowing them to create new outcomes that extend or alter institutionally prescribed goals while advancing those of individual users/citizens (Kimball, 2006).

Acknowledging citizens' varied roles as tactical information consumers and producers recognizes the significance of the ways in which the on-the-ground information and contextualized use come to be crucial parts of the broader knowledge base surrounding a technology, practice, activity, hobby, or knowledge site, defining and redefining products and activities while also enabling like-minded others to sidestep institutional or corporate strategies in similar ways, helping to bridge gaps between what is offered by institutions and what is desired by user-citizens. Harkening back to Robert Johnson's (1998) rhetorical analysis of the opposition between "system-centered" and "user-centered" views of technology development, such considerations move beyond questions of minor features or quibbles about cosmetic preferences to bring into focus fundamental concerns about purpose, control, and power. Tactical technical communicator citizens, then, often seek to remediate or redress these concerns at a local level, working at the fringes in order to tailor and/or improve institutions' prescribed strategic outcomes. And while this work may be motivated by individual user-citizens looking to meet their own immediate, subjective needs amid insufficient institutional strategies, as tactics move into community and dialogue with other similarly motivated citizen-users, they can coalesce, not only filling an isolated information gap but also critiquing and reconceiving the broader institutional mission and how it is executed vis-à-vis its users.

With regard to the institution of policing, citizens inhabit the role of "users" and find themselves working within and, at times, against the broader set of strategies that police use to order society. Their on-the-ground relationship with policy and practice moves them, at times, to find their own ways of engaging with the institution of policing. While some citizens voice their concerns about policing in more traditional forms of protest and activism, others engage tactically, sometimes in overlapping ways, to create microreforms or local changes by and through citizen influence gained through engagement with the technical information of policing. Before proceeding, a brief account of some key periods and movements in the history of American policing can offer context for understanding the relationship that some citizens have with this public institution and its strategies—strategies that themselves make available the tactical interconnection between the second and first levels that de Certeau (1984) describes.

American policing traces its roots, for the most part, to the English model of policing advanced by Robert Peel in the early part of the 19th century in England (Walker & Katz, 2007). Prior to this time, law enforcement in colonial America was informal and skeletal, with the "night watch" keeping an eye on property, participating in slave patrols, and monitoring disturbances of the peace. For the larger part of the 19th century, policing was deeply political and characterized by corruption and grift, machine politics, favors, and little consistency (Walker & Katz, 2007).

In the early 20th century, American policing moved from what Alpert and Dunham (2003) call the era of "no regulation" to the era of "self-regulation." As public concerns over enduring corruption matured, the early stages of the hierarchical paramilitary organizational structure that has come to characterize contemporary policing came into view. The subsequent era of "external regulation" in the latter part of the 20th century seems to be a culmination of larger cultural imperatives that emerged during the turbulent 1960s when large-scale war and civil rights protests pitted police against citizens, prompting the emergence of the first use of force policies in law-enforcement agencies (Fridell, 2010; Fyfe, 1979). The second half of the century saw professional organizations emerge, along with evolving best practices and more sophisticated, routinized training designed to improve preparation and regularize elements of policing based on case law and best practices (Alpert & Dunham, 2003, p. 9).

More recently, the rise of the Black Lives Matter movement, spurred in part by documented cases of police brutality toward Black citizens, as well as a growing sense that the ever-expanding mission of policing may be impractical and unsustainable, have prompted some to call for the defunding or even abolition of police. The latter movement has been driven, in part, by policing's aforementioned historical roots in slave patrols and the attendant belief that policing and the justice system retain racist vestiges rendering them incompatible with a vision for true justice, especially with respect to Black Americans. At the center of these concerns is the use of force, equitable enforcement, and accountability for how force is used. Such accountability remains elusive. While in recent decades, increased efforts have been made to bridge the citizen/police divide and nurture relationships through community policing initiatives and public outreach, there remains a persistent cultural belief among police that the circumstances under which policing happens are entirely unique, leaving citizens incapable of understanding or assessing police behaviors (Skolnick, 1993).

Use of Force and Institutional Strategy

A key site of contention regarding policing's role in American society is the intersection of the use of force and police discretion to make decisions about how and when to deploy force. While much of policing's history has bent toward professionalizing, regulating, and limiting officer behavior through training, policy, and best practices, discretion remains crucial to the work, with officer judgment difficult for citizens to challenge or fully understand (Davis, 1975; Fyfe, 1979). In order to appreciate the tactical work of copwatching citizens, it is important to situate copwatching within the context of institutional strategies used by the state to structure the work of policing, fortify policing as an institution, and protect its work, shielding sensitive information and officer discretion from public scrutiny and, consequently, accountability measures. Key policies and court cases offer shape and contour for the work of policing—how it is to be executed and documented, and who might legitimately participate in its execution and documentation (Walker, 1993).

Police and the state use an array of strategies comprised of a patchwork of laws, constitutional rights, and policies in order to preserve and protect their discretion regarding the use of force—in essence, officers' right to assemble information and then use it as they see fit to guide action. The effect of these strategies is, ostensibly, to preserve and empower police to make decisions, as well as to preserve the capacity to document and control narratives when police use force. The combination of police's discretionary action, their authorship of arrest narratives, and in-house disciplinary proceedings makes it difficult for citizens to "see" policing's work, which then makes it difficult for citizens to seek accountability.

Understanding the nature of policing strategies helps to clarify the reasons for citizens' tactical work while revealing, too, the gaps wherein citizen tactics might emerge. One good place to start is with qualified immunity. Rooted in a series of Supreme Court decisions over the past 150 years, qualified immunity emerged in response to a post–Civil War judicial doctrine (section 1983) supporting citizens' rights to sue government officials for violating the Constitution and citizens' civil rights. *Harlow v. Fitzgerald* established the current standard wherein officials are protected from prosecution in cases where conduct has not been deemed demonstrably unconstitutional by precedent (Williams, 2012). As currently recognized, qualified immunity, according to Williams (2012), "is a two-step inquiry: a court will dismiss a defendant from the suit if she did not violate the plaintiff's constitutional right or if the constitutional right she did in fact violate was not clearly established at the time of the violation" (p. 1298). In short, qualified immunity adds to the burden of proof for the plaintiff by raising the standard for precedent regarding the constitutionality or unconstitutionality of an action as "clearly established" precedents can be sufficiently nebulous to make civil charges difficult to sustain. Ideally, this has the effect of shielding public officials from spurious charges while still recognizing citizens' rights to seek accountability for officials' actions. With regard to policing, however, the heightened standards of qualified immunity can sometimes protect officers from legal consequences for uses of force that may, at times, seem puzzling to the public.

Beyond qualified immunity, certain policing-specific case precedents protect police discretion vis-à-vis the use of force from public accountability. In *Graham v. Connor* (1989), the Supreme Court held that all claims of excessive force by police officers must be evaluated using the Fourth Amendment's "objective-reasonableness" standard applied to officer behavior and reaffirmed the importance of the "totality of the circumstances" as necessary to understand and judge the appropriateness of contextual officer behavior. Importantly, the "reasonableness" of a particular use of force must be judged from the perspective of a reasonable officer on the scene, rather than with the 20/20 vision of hindsight.

These protections for officers' judgment function, in part, as a key component of the legal foundation for use-of-force policy, the agency-specific internal policy that establishes guidelines and guardrails for officers' use of force in each municipality or jurisdiction. Shaped in part by both statutory limits on force rooted in *Graham v. Connor* and the discretion and flexibility promised by qualified immunity, such policy empowers officers to use their discretion to choose force options commensurate with the threat level presented as they interpret them. Moreover, use-of-force policies establish the disciplinary protocols activated when deadly force is used; these protocols typically privilege internal review of force events, further shielding the public from information that might cast a negative light on policing while making it difficult to hold officers accountable for the use of force by shielding it from public scrutiny.

In combination, then, case law and policy help to codify the institutional strategies that secure policing's practices, as well as policing's distance from the public. By enshrining "reasonableness" as something possessed by officers and by affirming the centrality of officer discretion in use of force decisions, these intersecting and interpenetrating texts help establish a strategic apparatus that coordinates how policing works and how closely the public might engage with it (Clark, 1975). They render citizen judgment of police activity always/already problematic by demarcating which information matters in the use of force, who documents it, and who can access and act upon that information. The effect, ultimately, is that some citizens—especially marginalized citizens—often feel powerless, voiceless, and resigned vis-à-vis policing.

Copwatching and Tactical Citizen Actions

The public institution of policing, then, invites tactical action from citizens who seek agency vis-à-vis police accountability with regard to the use of force. With respect to policing, some citizens seek to define de Certeau's (1984) "second level" within the terrain of institutional strategy and control by opportunistically accessing the physical and legal spaces left available by institutional strategies in order to do tactical work as "copwatchers." In doing so, they attempt to bear upon how policing information is invented, used, and circulated in order to shape police practice and enhance accountability.

Brief History

Copwatching's origins trace to the 1960s and are oftentimes associated with the Black Panthers' program of police monitoring enacted primarily in Oakland, California (Toch, 2012; Nelson, 2016). Modern copwatching practice traces its roots to the early 1990s formation of Berkeley Copwatch, a local collective that has continued to influence the tactics and methods used by copwatchers and copwatching organizations across the country (Boyle, 2020). Markedly different from chance recording of controversial police actions (such as George Holliday's recording of LAPD officers beating Rodney King), copwatching as an organized practice centers on routinize, structured monitoring of police activity. Depending on the level of sophistication and organization, copwatching can also include a suite of supporting activities, such as offering training and support for other citizens or advocacy groups, protesting, curating and storing video data for citizen or attorney use, and leading local initiatives with respect to policing and policy (see, for instance, Justice Committee, 2021; Portland Copwatch, 2021).

Observed Practices and Interviews

As part of my research into copwatching as a tactical practice, I studied the copwatching activities of three different organizations in three different cities and interviewed ten different copwatchers across the same three cities. One organization is on the West Coast (city 1), and two organizations are located in Midwestern cities (cities 2 and 3, respectively). Some of these voices are integrated later in this section. Names of cities and individual copwatchers are protected per IRB protocol in order to preserve anonymity and encourage candor.

While the above description of copwatching emphasizes observation and recording as central activities, again, a range of tactics constellate to comprise the broader work of copwatching. Here, I spotlight four overlapping tactical practices/tactical sites wherein copwatchers engage with officers' protected, enshrouded decision making and repurpose information surrounding the act of policing, particularly the use of force. For each, I attempt to establish how copwatchers find spaces within state and policing strategies where they can impact policing in pursuit of their own goals or the goals of other citizens within a community. While I witnessed or discussed each of these tactics through interviews and observation, I attempt here to connect those encounters with more general copwatching practices beyond these three cities.

Sousveillance

Monitoring police and documenting activity, typically through use of digital video cameras or smartphones, becomes part of a broader form of sousveillance, which Mann (2013) translates, roughly, as "watching from below."

Copwatching citizens reverse the directionality of the modern state's surveillance strategy by inviting citizens to turn their cameras on the state—on to police officers—to track their behaviors, words, and activities and to use this digital evidence to enforce accountability.

From the standpoint of tactics, the digital camera used in this fashion enables citizens to gather information that is typically out of view and to possess information about policing that has historically been unavailable to citizens. This "legal evidence," as a city 1 copwatcher notes, might "exonerate someone or hold a cop accountable." As another copwatcher from city 1 offers, "There's something about video that's just undeniable. Like, when you see a police officer beating someone, there's no, like, 'Oh, is he really beating someone? . . . There's something about the power of video that other mediums of information don't have."

Know-Your-Rights Training

The power of police vis-à-vis the citizenry lies in police officers' sanctioned capacity to use force (Bittner, 1971). However, at times, police leverage that capacity, in part, through citizens' civic ignorance, insinuating that compliance is expected, even when it is not legally required. Additionally, many citizens do not know that they can legally record police activity in most states. Informed by ACLU programs and trainings, organized copwatching groups often provide citizens with "Know Your Rights" trainings, particularly in areas with heavy police presence (ACLU, 2021). These trainings offer insight into how to protect oneself, how to handle conflict and violation of rights, and what needs to happen when citizens are pulled over or when police are at the door. Copwatchers, then, provide the kind of civics information and constitutional knowledge needed for citizens to understand and act upon their rights, as well as understand the limits on those rights, revealing gaps in state strategy.

Agonistic Involvement in Police-Citizen Interaction

From a technical communication perspective, monitoring and documenting police activity brings critical information from the arrest into the light: typically happening out of view, police work, when documented by citizens, can be witnessed and scrutinized, with video footage widely circulated to members of the public or used as evidence in a court of law, if need be.

But as Simonson (2016) has argued, citing Chantal Mouffe's theory of agonistic political engagement, copwatchers also alter the very chemistry of a citizen encounter with police. In a form of "agonistic" engagement, copwatchers move beyond mere observers, with copwatching becoming a way "to provide meaningful *input into* the workings of that system" (p. 397). Copwatchers' physical presence, the power of the camera, and their direct engagement with officers have the capacity to alter officer behavior by infusing, as Simonson (2016) argues, citizens' constitutional interpretation directly into the police officers' interactions with a citizen. One copwatch organization leader I observed in city 2, for instance, would pointedly request information from officers and openly question officer actions. If we understand the arrest narrative as one of the most crucial forms of technical documentation (see Seawright, 2017 on police reports) in policing, citizen copwatchers participate in the construction of that narrative through their presence, actions, and words in the midst of the police-citizen interaction. One copwatcher from city 3 referenced such influence: "[T]o me, a very good interaction (with police) is if somebody gets pulled over . . . someone shows up with a camera, and the police officer doesn't want to be on camera and just lets them go. That's great!"

Database Development

A final tactic that some organized copwatching groups enlist is database development. Modeled on activist organization WITNESS's human rights-related video archiving practices, databases are used by some copwatching organizations to archive and organize collected footage of police activity. Such archiving enables copwatching organizations to better utilize digital assets by tracking patterns of officer behavior and providing needed footage to citizens and attorneys who seek out evidence (Berkeley Copwatch, 2020). In describing its organization's database project, El Grito de Sunset Park (2021) highlights the collective power of the video footage the organization has captured: "[I]n aggregate, these videos have the potential to help expose patterns of abuse, surveillance and harassment that many communities of color experience on a regular basis" (para. 5).

For citizens and copwatchers, these databases, still in their early stages of development and few in number, offer opportunities to access and use specialized policing information—the *content* of an arrest narrative—for their own purposes. Footage can be saved, shared, and used as evidence in court. Historically, again, police report narratives give officers the near-exclusive ability to tell the story of an encounter and rationalize their behaviors vis-à-vis citizens with little citizen recourse. The database, though limited, offers some citizens the rhetorical means to generate arguments that could either promote officer accountability or prove exculpatory for citizens if they are being charged with a crime.

Copwatching as Tactical Technical Communication: Operating in the Gaps Between Strategies

By virtue of their institutionalized power and individual discretion, police officers are frequently positioned as executors of the law, their subjective responses to situational contexts and citizen behaviors unavailable to citizens. This protected discretion becomes part of a broader institutional strategy that places the officer and, by extension, policing itself largely outside of judgment, in part because citizens cannot access the officer's situated behavior, which can only be understood—incompletely—by those able to understand the "reasonable-officer" subject position.

Copwatchers, I argue, are tactical technical communicators engaged in what de Certeau (1984) calls the "art of the weak"—tactical in their measured assessment of the limitations of institutional strategies and their own ability to find sites and "opportunities" both inside and outside of those strategies to tailor policing and enact change, oftentimes at the microlevel, to reposition, educate, and alter outcomes. Importantly, this work is embedded in the specialized discourse and the information of policing. Copwatchers, as de Certeau (1984) notes of tactical citizens, "must accept the chance offerings of the moment" (p. 37) and build literacies, coalitions, and knowledges that enable them and other citizens to leverage the affordances of a "moment" through data gathering and informed interventions. Indeed, copwatching's chief goal, it seems, is to impact the very invention of police interaction with citizens within the space of detainment wherein officer discretion and interpretation have historically gone unchallenged. Copwatchers do this by cleverly leveraging their own constitutional protections in an almost ironic setting—concurrently, as the power of the state is enacted on and against individual citizens through constitutionally protected policing (Regan, 2015). Such monitoring offers access to key information and opportunity to condition the invention of that information—the unfolding arrest narrative—troubling, even briefly, the strategic conditions that protect officer discretion from public view and critique.

In *The Practice of Everyday Life*, de Certeau (1984) describes "operational schemas" put into place by the state that regulate and order at one level but that are subject to remaking and recombining on the ground, becoming a "second level" of sorts that exists both within and without the dominant order and complex of strategies. Here, citizens seek leverage not through head-to-head confrontation with power but, rather, through individual citizens' efforts to find, frame, reframe, cobble, and make anew the conditions of their lives to best suit their individual goals and realize their desires. As scholars in technical communication have noted, such efforts frequently center on information and questions of type, access, circulation, collaboration, and more. To return to Holladay's (2017) and Bellwoar's (2012) respective studies, it is clear, again, that citizens sometimes move tactically to apply knowledge, occupy gaps, and leverage information when they perceive that an institution falls short. The stakes may be extremely high (e.g., Bellwoar's [2012] pregnancy study) or comparatively low (e.g., Reardon et al.'s [2017] study of gamers). Regardless, tactical technical communicators—citizens working in and against the framework of institutions—find a different way to pursue and achieve the outcome they seek, rather than meekly submit to the institution's power and strategies.

For copwatchers, the potency of this tactical action begins with a disposition toward sousveillance that recognizes citizens' weakness relative to the state but finds power in constitutionally protected, citizen-sponsored observation "from below." As citizens, they work to gather and control vital information—about participants, the scene, precipitating events, sequence of events, and speech—that can be used by citizens in various ways to pursue their ends and claim some power, both for themselves and for the detained citizen(s). Copwatching organizations, then, focus on invention strategies geared toward developing new texts with new rhetorical possibilities rooted in different uses of meaningful information (e.g., archiving and circulating video via databases and social media, respectively). They ask, in the face of institutional power and structural constraint, "What is left?" and "What can still be done?" isolating ways of legally involving themselves in the development and exchange of police-related information that otherwise oftentimes remains out of view.

Moreover, copwatching's broader set of tactics bear upon the work of policing and condition public views/sentiment toward policing. As Andrea Pritchett, cofounder of Berkeley Copwatch, notes: "There's something fundamentally shifting about our project of cop watching. It doesn't have to just be about watching. Now it's about organizing and what we do with the information" (Boyle, 2020, para. 4). In their close attention to mastering the scope of rights and constitutional protections that they as police monitors and general citizens have vis-à-vis police, copwatchers become what

Kimball (2017, p. 2) calls "limited-scope experts," enabling them to focus attention on affordances that enable citizens to hold police accountable with or without the direct support of the police or the state.

Even more, the approach to policing that copwatchers take on during many encounters with police as they directly engage with citizens seeks to impact the very character of that information by playing a role in its construction. In doing so, copwatchers seek to change outcomes through information and how it is invented, formulated, and constellated. As Simonson (2016) argues, copwatchers take on an agonistic stance when they are physically present and actively documenting a citizen-officer interaction. This means that they maintain a relationship with the institution to resist or alter it, rather than simply capitulate to it or reject it. In doing so, copwatchers involve themselves to some extent in officer action and discretion and, thus, what the footage will ultimately reveal, through impacting the "embodied narrative" of the police-citizen interaction (Bock, 2016). Simonson (2016) notes that this is nothing short of inserting local community values and interpretation of constitutional rights as citizens. Rather than merely submitting to the will of police, copwatchers participate in an interpretive act, adding information to the police-citizen interaction, while adding, too, a measure of peer pressure or behavioral conditioning.

To return to the de Certeau (1984) quote above regarding the "art of the weak," tactics are "ways of operating" that "intervene in a field which regulates" citizens (p. 30). Copwatchers seek out the opportunities made available by the strategies used to regulate the citizenry from law enforcement's perspective. The "weak" vis-à-vis policing take advantage of protections on speech and powerful new media to use information—in this case, video footage—in ways that they see fit, rather than how the institution would prefer. Indeed, as policing has slowly embraced body and dashcam video footage as a tool to exonerate (a strategy deeply enmeshed in power relations and the privilege of releasing footage on departments' own terms), copwatchers struggle to co-opt the same "information"—the theoretically objective "text" of an arrest—possessing and deploying it to push for accountability. In doing so, copwatchers are able to move closer to participating in the frequently sealed-off space for officer discretion that *Graham v. Connor* preserves, something of a final frontier for accountability advocates, and "radically sharing" this information to engage with publics and promote broader discourse about policing practices.

In sum, the space of the police-citizen interaction—really, policing's nearly-exclusive narrative space—comes to be refigured by tactical citizens who are frequently dissatisfied with its outcomes. By isolating spaces for opportunity, copwatchers, through presence and interaction, seek to alter the unfolding event; by capturing the events using digital video and constitutional affordances/protections, they claim a text that they can then use, modify, circulate, amplify, and so on. to pursue their vision of how policing should look in its execution. While broad, systemic change may be beyond reach and the strategies of the state subject to larger forces beyond citizen control, smaller acts, using available materials, offer potential for intervention. Echoing Kimball (2006), copwatchers tactically assert "here is what we want" in response to the state's claim that "here is how it must be done" (p. 74).

References

Alpert, G., & Dunham, R. (2004). *Understanding police use of force: Officers, suspects, and reciprocity.* Cambridge University Press.

American Civil Liberties Union. (2021). *Know your rights: Stopped by the police.* ACLU https://www.aclu.org/know-your-rights/stopped-by-police/

Bellwoar, H. (2012). Everyday matters: Reception and use as productive design of health-related texts. *Technical Communication Quarterly, 21*(4), 325–345. doi.org/10.1080/10572252.2012.702533

Berkeley Copwatch (2020). *People's database.* Berkeley Copwatch. https://www.berkeleycopwatch.org/people-s-database

Bittner, E. (1972). *The functions of the police in modern society: A review of background factors, current practices, and possible role models.* National Institute of Mental Health.

Bock, M. A. (2016). Film the police! Cop-watching and its embodied narratives. *Journal of Communication, 66*(1), 13–34. doi-org.libproxy.uwyo.edu/10.1111/jcom.12204

Boyle, M. (2020). *Interest in community police watch training soars as courses go online.* San Francisco Public Press. https://www.sfpublicpress.org/interest-in-community-police-watch-training-soars-as-courses-go-online/

Davis, K. (1975). *Police discretion.* West.

de Certeau, M. (1984). *The practice of everyday life.* University of California Press.

Ding, H. (2009). Rhetorics of alternative media in an emerging epidemic: SARS, censorship, and extra-institutional risk communication. *Technical Communication Quarterly, 18*(4), 327–350. doi: 10.1080/10572250903149548

El Grito de Sunset Park. (2021). About the project. https://elgrito.witness.org/about-the-project/

Finer, B. S. (2016). The rhetoric of previving: Blogging the breast cancer gene. *Rhetoric Review, 35*(2), 176–188. Doi.org/10.1080/07350198.2016.1142855

Fridell, L. (2010). Deadly force policy and practice: The forces of change. In Candace McCoy (Ed.), *Holding police accountable.* (pp. 29–54). Rowman & Littlefield.

Fyfe, J. (1979). Administrative interventions on police shooting discretion: An empirical examination. *Journal of Criminal Justice, 7*(4), 309–24. https://doi.org/10.1016/0047-2352(79)90065-5

Graham v. Connor. 490 U.S. 368. (1989).

Holliday, D. (2017). Classified conversations: Psychiatry and tactical technical communication in online spaces. *Technical Communication Quarterly, 26*(1), 8–24. doi.org/10.1080/10572252.2016.1257744

Johnson, R. (1998). *User-centered technology: A theory for computers and other mundane artifacts.* State University of New York Press.

Justice Committee. (2021). CopWatch. https://www.justicecommittee.org/cop-watch

Kimball, M. (2006). Cars, culture, and tactical technical communication. *Technical Communication Quarterly, 15*(1), 67–86. doi.org/10.1207/s15427625tcq1501_6

Kimball, M. (2017). Tactical technical communication. *Technical Communication Quarterly, 26*(1), 1–7. doi.org/10.1080/10572252.2017.1259428

Mann, S. (2013). Vigilance and reciprocal transparency: Surveillance versus sousveillance, AR glass, lifelogging, and wearable computing. *2013 IEEE International Symposium on Technology and Society (ISTAS): Social Implications of Wearable Computing and Augmented Reality in Everyday Life, 2013*, 1–12. doi: 10.1109/ISTAS.2013.6613094

Nelson, S. (Director). (2016). *The Black Panthers: Vanguard of the revolution* [Film]. PBS.

Petersen, E. (2018). Female practitioners' advocacy and activism: Using technical communication for social justice goals. In Godwin Y. Agboka and Natalia Matveeva (Eds.), *Citizenship and advocacy in technical communication: Scholarly and pedagogical perspectives* (pp. 3–22). Routledge.

Pflugfelder, E. (2017). Reddit's "explain like I'm five": Technical descriptions in the wild. *Technical Communication Quarterly, 26*(1), 25–41. doi.org/10.1080/10572252.2016.1257741

Portland Copwatch. (2021). *Welcome to the Portland Copwatch web page.* Portland Copwatch. http://www.portlandcopwatch.org/

President's Task Force on 21st Century Policing. (2015). *Final report of the president's task force on 21st century policing.* https://cops.usdoj.gov/pdf/taskforce/taskforce_finalreport.pdfreport

Reardon, D., Wright, D., and Malone, E. (2017). Quest for the happy ending to Mass Effect 3: The challenges of cocreation with consumers in a post-Certeauian age. *Technical Communication Quarterly, 26*(1), 42–58. Doi.org/10.1080/10572252.2016.1257742

Regan, L. (2015). *Policing the police: Your right to record law enforcement.* Civil Liberties Defense Center. https://cldc.org/policing-the-police

Sarat-St. Peter, H. (2017). "Make a bomb in the kitchen of your mom": Jihadist tactical technical communication and the everyday practice of cooking, *Technical Communication Quarterly, 26*(1), 76–91, doi.org/10.1080/10572252.2016.1275862

Seawright, L. (2017). *Genre of power: Police report writers and readers in the justice system*. NCTE.

Simonson, J. (2016). Copwatching. *California Law Review, 104*(2), 391–445. http://www.jstor.org/stable/24758728

Skolnick, J. (1993). *Above the law: Police and the excessive use of force*. Free Press.

Toch, H. (2012). *Cop watch: Spectators, social media, and police reform*. American Psychological Association.

Van Ittersum, D. (2014). Craft and narrative in DIY instructions. *Technical Communication Quarterly, 23*(3), 227–246. doi.org/10.1080/10572252.2013.798466

Walker, S. (1993). *Taming the system: The control of discretion in criminal justice, 1950–1990*. Oxford University Press.

Walker, S., & Katz, C. (2010). *The police in America: An introduction* (Seventh edition). McGraw-Hill.

Williams, J. (2012). Qualifying qualified immunity. *Vanderbilt Law Review, 65*(4), 1295–1336.

Chapter Seven

Tactical Tech Comm and Failures in Crowdsourcing Mass Production

John T. Sherrill

In this chapter, I argue that the Pussyhat Project applied a logic of mass production to a situation that called for mass customization. In contrast, the Trump administration's response to the COVID-19 pandemic applied a logic of mass customization to a situation that called for mass production. This logic, in large part, helped shift the responsibility of public health away from the Trump administration onto the American public by framing mask wearing and mask making as individual choices (under the guise of freedom). I situate these two contrastive cases as examples of partially failed tactical technical communication. I compare these two cases in particular because they both address creating and distributing goods at a national level, center on issues of public health as well as individual identity and collective rights, and because they bookend the Trump presidency as examples of wearable craftivism.

The examples throughout this chapter correspond to national tensions between morality and personal freedom as well as the logic of production underlying them. In order to understand and analyze the logics of production in each example, I draw from definitions of mass customization and postindustrial production in multiple fields, while situating these examples as craftivist tactical technical communication. Using Randall and colleagues' *Principles for User Design of Customized Products* (2005), I suggest ways that the Pussyhat Project could have incorporated a logic of mass customization and argue that it is important for technical communicators to recognize these underlying logics in order to help encourage ethical user choices and appropriate production methods as the situation calls for.

Craftivism as Tactical Technical Communication

Drawing from Kimball's (2017) definition of tactical technical communication, I situate craftivism in this chapter as a form of tactical technical communication. Craftivism often involves providing instructions for how to make a particular object rather than educating crafters on "how to become," distributing patterns and designs for DIY projects via the internet ("radical sharing"), and developing and sharing kits and documentation. Further, craftivism is often a form of resistance, aiming to enact social or political change in response to institutional or structural problems and thus echoes the "social-justice turn" described by Moore and others (2019). Craftivism combines craft and activism, using craft to enact social and/or political change. In 2008, Betsy Greer defined craftivism as being "dialogic, small scale, and view[ing] the 'act of "making"' as part of the dialogic process" (Greer, 2008; Sherrill, 2019, p. 18). Sarah Corbett expanded on Greer's definition, situating craftivism as "slow activism" that encourages reflection and dialog because crafting requires time and attention, and Corbett argues that craftivism should be both public and "small, attractive and unthreatening" to initiate dialogue through the visibility of crafting, ideally with a group (Corbett, 2013, pp. 5–6). Given that mass customization shifts part of the responsibility of the design process onto users, it makes sense to examine DIY communities as a source for examples and analysis of tactical technical communication. Further, craftivist projects foreground the ethical and moral aspects of DIY design and decision making, much like mass customization does implicitly. One of the most visible and well-known recent examples of craftivism is the Pussyhat Project.

The Pussyhat Project

In 2017, as part of the Women's March on Washington in response to Donald Trump's election, thousands of participants wore pink, "eared" hats, many of them hand crafted from patterns distributed by the Pussyhat

Project in 2016. According to the creators of the Pussyhat design, Krista Suh and Jayna Zweiman, the hats were intended to create a visual "sea of pink" (Pussyhat Project, n.d.-a) and to serve as a "unified statement" (Kahn, 2017). However, the design and marketing of the hats received criticism on multiple fronts, namely, that they were exclusionary because they reduced women's identities to their genitals (and privileged cis-gender women), were white-centric, and homogenously represented women, alongside criticisms of the broader Women's March on Washington (Derr, 2017; Huber, 2016; Dejean, 2016; Compton, 2017).

Although Suh and Zweiman responded to critiques of the Pussyhat Project, the project did not actively promote or solicit alternative designs, such as Rachel Sharp's trans-pride hat or Jacqueline Cieslak's Black Lives Matter pattern (WrrrdNrrrdGrrrl, 2017; Cieslak, 2017). A FAQ section on the Pussyhat Project website explained that alternative designs were acceptable as long as they were pink (Pussyhat Project, n.d.-b). At best, the FAQ page on the site acknowledged that some crafters would want to "use a different pattern" or use methods other than knitting (Pussyhat Project, n.d.-b) but did not actively promote customization or highlight custom designs via social media or official pattern pages. Given this, as well as the criticisms of the design for being exclusionary, I argue that despite making use of crowdsourced labor, the Pussyhat Project did not take full advantage of the affordances of crafting communities and was still rooted in a logic of industrial mass production rather than postindustrial mass customization via crowdsourcing. In other words, Suh and Zweiman envisioned crowdsourcing as a means of DIY mass production rather than an opportunity to generate a wide range of designs and customized versions; something to be permitted rather than being a feature or the point. From this perspective, although the activist ends and political goals of the Pussyhat Project were radically different from those represented by mass-produced red MAGA hats, the underlying logic of a one-size-fits-all approach to mass producing hats led to many groups feeling excluded and unrepresented by the uniformity of the official Pussyhat designs.

As mentioned in the CFP for this edited collection, tactics are "a means of taking action *now* under imperfect conditions instead of speculating what action one might someday take in 'the realm of fiction and future,'" and the Pussyhat Project's imperfections have been critiqued at length in the sources cited above. With this in mind, the following sections describe a logic of mass customization and principles for designers, technical communicators, and activists looking to design systems and tactical responses that facilitate mass customization. To help illustrate some of these principles, I map them onto the Pussyhat Project as a frame of reference. In the last section of this chapter, I briefly argue that in some cases (e.g., pandemic mask production), a logic of mass production and one size fits all is still the most fitting approach.

Mass Customization

Mass customization is a familiar part of everyday life for many around the world, particularly in the United States. Ordering a shirt, with carefully selected fabric from a dropdown menu, custom embroidery, and a fit that needs minimal tailoring thanks to detailed measurements and user feedback, is a common experience. Or perhaps purchasing a new car, similarly, customized via a series of checkboxes and virtual previews, manufactured and delivered to the nearest dealer within weeks. Or simply lunch: a sandwich personalized through a few check boxes in an app (no cheese, extra pickles and lettuce, and a note requesting that it be cut in half) delivered via a localized version of Uber—just one among many other customized sandwiches on the driver's delivery route for that hour. In simple terms, mass customization is a shift from 'one-size-fits-all' mass production to custom fit at a mass scale via automation and routinized labor (as opposed to haute couture or artisanal production), corresponding with the shift from industrial to postindustrial production.

Though there is no definitive transition from industrial to postindustrial, the transition is generally associated with increased digital automation. Daniel Cohen associates this transition with the software industry, specifically, the application of software-development principles to the development and production of other goods, namely, that "It costs a lot to conceive a piece of software, but it does not cost much to manufacture it" (Cohen, 2008, p. 5). That is, there is a large investment of mental labor in the design process, but once sufficient infrastructure is in place, reproducing copies or variants is relatively inexpensive. Although practices of mass customization are relatively old, "anticipated ['as a technological capability'] in 1970 by Alvin Toffler in *Future Shock* and delineated (as well as named) in 1987 by Stan Davis in *Future Perfect*" (Pine, 1993, p. xiii), recent advances in digital fabrication and additive manufacturing technologies, combined with a shift towards what Nick Srnicek (2016) terms platform capitalism, have led to further proliferation of mass customization as an operating model. In practice, this means that for companies like Ford or Subaru, the difference in cost between producing 10,000 identical vehicles and 10,000 unique vehicles is marginal, with significant differences in customer satisfaction.

Joseph B. Pine (1993) summarized differences between mass production and mass customization succinctly, writing, "While the practitioners of Mass Production share the common goal of developing, producing, marketing, and delivering goods and services at prices low enough that nearly everyone can afford them, practitioners of Mass Customization share the goal of developing, producing, marketing, and delivering affordable goods and services with enough variety and customization that nearly everyone finds exactly what they want" (p. 44). This is accomplished affordably "through economies of *scope*—the application of a single process to produce a greater variety of products or services more cheaply and quickly" (p. 48). To simplify further, complexity is free.

Originating in the additive manufacturing industry and consumer 3D-printing communities, the phrase *complexity is free* suggests that aside from the potential time investment during the design process, fabricating complex designs is no more expensive (and often not necessarily any harder) than fabricating a simple design (Deloitte, 2014). Though the origins of this phrase lie in 3D printing technology, in practice, the same principle applies to mass customizing patterns for pussyhats and masks fabricated via crowdsourcing. As explained in a Deloitte Insights post (2014), "When people say complexity is free, they are implicitly recognizing that AM [additive manufacturing] technology can be incredibly versatile. Because it generally produces objects 'layer by layer,' it can fabricate items that simply cannot be produced using other means" (para. 1). Similarly, knitting and crochet both produce objects row by row (with the capacity to yield immensely complex designs), albeit with human hands rather than robots in most cases.

Regardless of the end product, a logic of mass customization recognizes that for practical purposes, physical objects can be produced, manipulated, and often behave in ways increasingly like digital media due to advances in additive manufacturing and other related technologies. Complexity is free because the complexity is first digital, often automated, and mundane to manufacture. As Neil Gershenfeld (2012) wrote, referring to advances in digital fabrication and additive manufacturing (e.g., 3D printing), "The revolution is not additive versus subtractive manufacturing; it is the ability to turn data into things and things into data" (p. 44). Like with editing text before printing, the design of an object can go through multiple iterations and revisions and still later be edited quickly after spotting a proofreading error in a test print. Of course, this process is as old as the technology of movable type, in the sense that moveable type is both modular and automated. What increasingly differentiates the present moment from the past is that such modularity and automation can be applied to a much wider range of materials and applications beyond 2-dimensional text and sheets of paper or images on screen.

A decade before Neil Gershenfeld, Lev Manovich (2002) outlined 5 principles for new (digital) media and noted parallel logics underlying digital media and postindustrial manufacturing/consumption. Most relevant to this chapter, Manovich (2002) also forecasted that the variability of digital media—which allows users to customize interfaces at their will—would have a broader social impact, arguing that "every choice responsible for giving a cultural object a unique identity can potentially remain always open," and in doing so, "making a choice involves a moral responsibility" as every constant becomes variable (p. 44).

This freedom of choice has two important implications: first, that "by passing on these choices to the user, the author also passes on the responsibility to represent the world and the human condition," and second, that this choosing becomes "moral anxiety" (Manovich, 2002, p. 44). Nearly 20 years later, we have witnessed the implications of shifting responsibility and moral anxiety at a national level very clearly throughout the presidency of Donald Trump. This shift has been foregrounded with the COVID-19 pandemic and antimask and antivaccine rhetoric, and I argue that a similar shift was present in critiques of the Pussyhat Project.

Simultaneously, in shifting the labor of customization and choice onto individual consumers, customization also theoretically affords individualized fit and partially fulfills the goals of user-centered design. But why should customers pay for doing additional labor of customization themselves? Although better fit is one reason and making the process as simple as possible promises better fit with minimal effort, Franke and colleagues (2009) suggest that there is also value in the process of customization itself.

User Investment in Customization

In *The "I Designed It Myself" Effect in Mass Customization*, Franke and colleagues (2009) analyzed the customization of "self-expressive and publicly consumed products," including "scarves, T-shirts, cell phone covers, skis, and watches" in order to better understand why consumers value self-designed products more than stock designs when using systems configured for mass customization (p. 138). Through a series of studies, the authors controlled for variables such as fit, perceptions of quality, and time investments to isolate the role of what they term "I designed it myself." They define this concept as "feelings of accomplishment arising from the process of self-designing" (p. 137). The initial study found that participants who designed their own T-shirt were willing to pay over 40% more for a shirt of their own design than they were for a mass-produced (but still fitting) shirt. The initial study did not account for other potential explanations though, such as the amount of time participants invested in the overall process. But, after further studies to isolate relevant variables, the results consistently suggested that the simple act of participating in designing a product makes people value the resulting product more than when purchasing a premade design. Further, that perceived value grows when participants are given greater control over the design processes. For example, uploading a photo or adjusting the layout of a design led to greater value than if users were restricted to choosing between colors. Franke and colleagues (2019) note that this relationship between value and agency is probably "not a linear function," though, as greater flexibility also increases the potential for a mediocre outcome and/or greater time investment and effort (p. 137). In other words, effectively using mass customization to maximize perceived value is a balancing act between the amount of effort required by users, design flexibility allowed, and fit, in relation to how much the user feels they contributed to the process.

In addition to giving users more control over the design process, Franke and colleagues (2009) suggest that "immediate (positive) feedback," either automatic or crowdsourced through user communities, could help support users throughout the customization process (p. 138). For educators who teach technical communication and/or incorporate design projects into their classes, striking a balance between scaffolding and control is likely already familiar, along with providing encouraging feedback. It is important to note here that unlike in a classroom setting, though, the goal of designing a product via customization is tactical in nature. In the classroom, instructors hope that students enter with the purpose of becoming/*being* designers, artists, scientists, engineers, and so on and that the education process helps students take on some form of professional identity (though certain classes are sometimes viewed only as a requirement for graduation). In contrast, one typically does not hope to *become* a fashion designer by uploading a custom T-shirt design or a professional athlete by customizing the look of Nikes as part of the checkout process (though *feeling* like a professional designer/athlete is certainly part of the appeal). Rather, for most consumer products, users are trying to achieve fit for a specific task and then move on once finished.

With definitions for digital media, craftivism, and mass customization in mind, in the following sections, I describe how technical communicators and designers can help ensure that mass customization is applied successfully in tactical situations, and for craftivist projects specifically. I first describe principles for designing interfaces to facilitate customization. I then map these principles onto the Pussyhat Project, highlighting opportunities where these principles were overlooked. I follow that discussion with an example of how a university pep band successfully uses mismatched hats to create a collective visual identity on a much smaller scale. I conclude with a brief discussion of DIY mask making during the COVID-19 pandemic and implications for technical communicators.

Principles for User Design of Customized Products

In 2005, Randall and colleagues (2005) published *Principles for User Design of Customized Products.* Many of the principles discussed are likely familiar to technical communicators working in industry, given the relative ubiquity of customization options in e-commerce and common apps. Randall and colleagues (2005) do not address mass customization by name, likely because the article addresses the user-facing side of the process rather than manufacturing and production, but the principles still apply across a range of situations. These principles are summarized in table 7.1.

Although the actions listed in table 7.1 were recommended based on "different user interfaces for customizing laptop computers," rather than open-source activist designs, the principles are still applicable to craftivist projects (Randall et al., 2005, p. 69). Generally, the Pussyhat Project website excelled at the following:

1. "Providing starting points." The website included dozens of patterns for knitting, crocheting, sewing, and looming hats for a range of expertise.
2. "Teaching the consumer" by providing a variety of additional resources for each crafting method, along with
3. "Exploiting prototypes to avoid surprises," more specifically "providing rich illustrations of the product" throughout. Many different examples were provided in the form of illustrations, videos, and images.

Table 7.1. Principles for User Design of Customized Products

Problem	Principle	Action
Some consumers have more knowledge about the product than others	Customize the customization process	Provide novice consumers with a needs-based interface Provide expert users with a parameter-based interface
Not all consumers are interested in fully exploiting the potential of customization	Provide starting points	Provide multiple access points for customization
Customizing a product is a cognitively challenging task typically requiring many iterations	Support incremental refinement	Allow consumers to bookmark their work Allow for side-by-side comparison Provide short-cuts through "attribute space"
Since customized products are tailored to a specific consumer, the consumer typically must order a product before having seen or tested it	Exploit prototypes to avoid surprises	Provide rich illustrations of the product Provide increasing levels of fidelity in prototypes as the customization process progresses
Consumers know very little about the options available to them as well as how these options are useful in fulfilling their needs	Teach the consumer	Provide "help buttons" leading to meaningful information Explain the product attributes and how they map to design parameters Show the distribution of design parameters and product attributes across the consumer population

Source: T. Randall, C. Terwiesch, & K. T. Ulrich (2005). Principles for User Design of Customized Products. *California Management Review, 47*(4), 68–85, 69. https://doi.org/10.2307/41166317

In this regard, the designers of the project anticipated a variety of means of production and facilitated the mass production of standardized hats via effective technical communication. However, the project website was less successful at "customizing the customization process" and "supporting incremental refinement." Though the website includes beginner, intermediate, and advanced patterns, the visual design of the hats remains largely the same. It seems the FAQ section of the website was updated in response to critiques of the original hat design, but this was not reflected in the officially endorsed patterns. That is, although the critiques of gender essentialism and colorism/racism were eventually reflected in the FAQ page (Pussyhat Project, n.d.-b), the underlying logic of mass producing standardized pink hats remained intact and did not reflect any change in the diversity of patterns or finished hats. In that regard, the degrees of freedom for customization were restricted for official purposes. Consequently, support for incremental refinement was also restricted within the officially endorsed channels and would likely echo existing patterns rather than those that diverged from the established standard. Similarly, though the website was partially successful at "teaching the consumer," the action of "showing the distribution of design parameters and product attributes across the consumer population" was not demonstrated on the site as readily as in critiques before and after the march. While the Pussyhat Project still largely succeeded at its intended purpose of providing a visual "sea of pink" (Pussyhat Project, n.d.-a), these limitations help demonstrate where the project could have been more inclusive by applying principles of mass customization, particularly given that it already made use of crowdsourcing from its origin, and given the longstanding history of DIY embellishment and customization associated with craft.

For craftivist approaches to activism, techniques such as crowdsourcing, generative design, and modular design all facilitate the creation of customizable and personalized designs to ensure individual fit en masse. In particular, craftivist projects involving textiles, text, or media that are primarily visual in nature allow individual craftivists to separate pattern generation from fabrication. That is, even for beginners, mastery of foundational stitches, seams, or forms can enable the completion of detailed and complex designs if a pattern already exists (for example, highly detailed pixel quilts consisting of only different-colored squares). Of course, following some patterns may require more advanced techniques. But like with software development, much of the initial labor is in creating a pattern that can be endlessly replicated digitally, relative to the labor of fabricating the design (which may be partially automated, depending on the project). To be clear, the labor of mass customization is not static or equal across all media, and it is important to recognize the role of materiality across the examples provided here. Any quilter knows that coming up with a pattern is still relatively less laborious than the overall process of cutting, piecing, pressing, assembling, quilting, and binding fabric into a finished quilt (though this manual labor can be outsourced, offsetting the actual labor necessary for consumers to own a custom quilt). Knitting a completed hat still takes more time and manual labor than sewing one with stretch fabric, or machine embroidering a stock hat, and still more time than comparable projects produced through digital fabrication (e.g., designs produced with 3D printers, laser cutters, vinyl cutters, printers, etc.). As demonstrated through the Pussyhat Project, inefficiency is partially the point of craftivism and creates opportunities for dialogic participation. But that should not mean ignoring the affordances of mass customization. In the following example, I describe how a university pep band uses a logic of mass customization to successfully engineer a collective visual identity while emphasizing individual fit. Doing so creates a cohesive community, albeit one whose identity can be appropriated with relative ease.

Consistent Visual Identity Through Diverse Hats

Although not known as a particularly feminist or craftivist organization, Michigan Technological University's Huskies Pep Band serves as an unlikely example of highly customized collective visual identity at a small scale and often employs craft. As a scramble band, the pep band does not march in step, and like similar bands, it is known for its comedic antics and satire (not unlike many activist groups). Fittingly, the band can generally be recognized by the collective presence of many people carrying instruments (though one could be easily confused, as the ensemble has included accordions, bagpipes, vuvuzela, various flaming brass instruments, an oven, and other unlikely/improvised noisemaking objects). Additionally, though the band generally wears black-and-gold-striped overalls, there are not always enough uniforms for all members. Further, multiple other university bands wear nearly identical overalls, so the uniform alone is not sufficiently distinct. Given this, one of the most recognizable collective features of the group, which also distinguishes it from other bands, is a multitude of mismatched hats.

Each pep-band member is required to wear a unique hat; the more eccentric, the better. This is significant as one cannot refuse to wear a hat, only to choose which hat to wear. Failure to do so allegedly results in a hat being applied via mob tactics, but creativity and individual identity are positively encouraged, often yielding customized and even handmade hats for all members. Despite the sheer variety of unique hats, collectively, the hats lend a distinct identity to the group as a whole, while embracing the individual identity of each member. Whereas a traditional marching band can be recognized by identical military-style shakos and matching mass-produced uniforms that deemphasize (or even strip) members of their unique identities, the Huskies Pep Band is recognized primarily by custom hats and to a lesser extent by mass-produced uniforms. It is important to note here that this choice of nonuniform visual identity is a tactical response to the status quo of military-style uniforms, like the pinkness of Pussyhats as a response to red MAGA hats. The Huskies Pep Band is distinctly recognizable from a visual standpoint precisely because other bands rely on uniformity with such consistency, and the use of custom hats can be easily appropriated—thus illustrating Manovich's argument (2002) that the ubiquity of customization makes the choice of what hat to wear a moral dilemma.

Along these lines, the Pussyhat Project was an attempt to create a collective visual identity that represented a moral choice in order to enact change through collective action. As the critiques mentioned throughout have established, it could have been more successful in representing a collective identity inclusively. At the same time, as Laura Portwood-Stacer (2007) pointed out in her critique of craftivism, individual acts of craft, no matter how politically charged, and individual acts of crafting (by choice rather than by traditional gender role), are not sufficient. Given this, mass customization can afford an important framework for realizing meaningful collective change while representing individual identities. It does not, however, guarantee meaningful collective action, given that mass customization originated in relation to concepts of just-in-time delivery, lean manufacturing, on-demand production, and minimizing/eliminating overproduction. That is, there is also potential for mass customization to reinforce individualized activist actions through a focus on the local or smaller groups rather than collective masses. Mass customization can produce thousands of stock pink pussyhats or red MAGA hats, thousands of unique hats, or a mix of both stock and personalized, regardless of politics or morality. Given this, it is not a fitting logic for all situations. In particular, this is demonstrated by the Trump administration's failure to follow a logic of mass production in relation to distributing masks in response to the COVID-19 pandemic.

To Wear or Not to Wear Versus Which to Wear

Throughout 2020, this failure was illustrated by the emergency production of varied DIY masks—at home, via volunteer networks, through Etsy, and so on—as well as emergency shifts in assembly lines designed for mass customization to mass produce masks. In that moment, a one-size-fits-all approach to the production of masks would have been a good thing, serving a basic medical function first and foremost and likely saving hundreds of thousands of lives. Further, the efficiencies that mass production affords help guarantee standard quality and greater effectiveness than homemade masks. However, disruptions to postindustrial supply chains have also foregrounded the limitations of just-in-time manufacturing and delivery along with the fragility of global supply chains based on a logic of mass customization, perhaps suggesting that such mass production in 2020 was less plausible than one would hope.

Regardless of whether a mask has been homemade or mass produced, and despite the diversity of colors and types, the simple act of wearing a mask has become a visual symbol of particular political beliefs within the United States. Trump's framing of mask wearing as a personal choice, rather than a public mandate, easily reinforced the narrative that choosing to wear a mask was indicative of "liberal" moral values and anti-Trump sentiment. One need only frame the issue as a matter of individual freedom and personal choice "to wear or not to wear" rather than "which to wear or else," to have devastating consequences when seemingly every choice is customizable. The initial production of masks by any means, whether DIY or mass produced, was absolutely necessary. But the proliferation of custom and nonstandard masks, following an underlying logic of mass customization, arguably only made it easier to reinforce the perception that wearing a mask was an issue of personal choice and expression of individual identity that possibly remains open, rather than an issue of collective public health. By the time mass-produced masks became readily available, perceptions of the collective identity that masks represented were already solidified. Although a ready supply of mass-produced masks would not have automatically countered Trump's persistent efforts to frame masking as an issue of personal choice, it would have provided some tactical resistance and greater individual and collective protection against COVID-19.

Given the respective successes and failures of the Pussyhat Project, mass-produced MAGA hats, and masks discussed throughout this chapter, it is important that technical communicators consider logics of mass customization and mass production, especially as part of the social-justice turn in technical communication. Tactical responses may not always afford the time necessary to standardize a mass-production process nor to devise interfaces or instructions for easy mass customization. It is therefore important that technical communicators be able to quickly recognize the logic of production that a given situation calls for in order to make the most of a given opportunity—both to minimize risks of excluding participants based on their individual (or collective) identities and to recognize when a one-size-fits-all solution can help encourage collective action or offer greater collective benefit than a customized approach.

Given Manovich's argument (2002) that shifting the responsibility of design choices onto users brings moral anxiety, Franke and colleagues' finding (2009) that users place value on things they designed themselves has important implications if there is a corresponding greater investment from a moral standpoint. That is, does "I designed it myself" translate not just into greater perceived value, but greater moral conviction or more deeply held beliefs? Is there a significant difference in conviction between someone who buys a hat off the shelf versus someone who designs their own? If so, technical communicators are in a position of considerable power and responsibility to tactically and strategically apply the I-designed-it-myself effect to influence participant investment in moral issues via methods of production. At the same time, it is important for technical communicators to recognize that the act of tactical customization does not mean that one "becomes" or takes on an identity long-term via temporarily acting to solve an immediate problem.

The examples throughout this chapter have focused on craftivism, which by definition involves dialog and making things. It is imbued with tensions between customization and mass production, individual and collective. Technical communicators are well positioned to take on, and learn from, craftivist work while addressing the questions and tensions presented above. We are already attuned to design methodologies and how to facilitate dialogic interactions between user-participants and designers, while considering who is represented, privileged, and impacted in participatory systems. When considering craftivism specifically, and any form of activism that requires the physical production of materials or objects, it's important that technical communicators consider the underlying logics of production and their political implications. Misalignment between activist/political goals and associated logics of production can have a significant impact on participants and users—particularly at a national level—and can make the difference between a tactical and a strategic response.

References

Cieslak, J. (n.d.). *PUSSYHAT—Black Lives Matter*. Stitch Fiddle. https://www.stitchfiddle.com/en/c/sj487n-bgkv74/quickview

Cohen, D. (2008). *Three lectures on post-industrial society* (W. McCuaig, Trans.). MIT Press.

Compton, J. (2017, February 7). *Pink "pussyhat" creator addresses criticism over name.* NBC News. http://www.nbcnews.com/feature/nbc-out/pink-pussyhat-creator-addresses-criticism-over-name-n717886

Corbett, S. (2013). *A little book of craftivism.* Cicada Books.

Dejean, A. (2016, November 12). *"Million women march" protest was appropriating black activism so organizers did this.* Fusion. http://fusion.net/story/369581/million-women-march-protest-appropriation/

Deloitte Insights. (2014, March 6). *3D printing: "Complexity is free" may be costly for some.* Deloitte Insights. https://www2.deloitte.com/us/en/insights/focus/3d-opportunity/3d-printing-complexity-is-free-may-be-costly-for-some.html

Derr, H. (2017, January 17). *Pink flag: What message do "pussy hats" really send?* Bitch Media. https://bitchmedia.org/article/pink-flag-what-message-do-pussy-hats-really-send

Franke, N., Schreier, M., & Kaiser, U. (2009). The "I designed it myself" effect in mass customization. *Management Science, 56*(1), 125–140. https://doi.org/10.1287/mnsc.1090.1077

Gershenfeld, N. (2012). How to make almost anything: The digital fabrication revolution. *Foreign Affairs, 91*(6), 43–57. http://www.jstor.org/stable/41720933

Greer, B. (2008). *Knitting for good! A guide to creating personal, social, and political change stitch by stitch* (Original edition). Roost Books.

Huber, C. (2016, December 2). *The problem with the women's march on Washington and white feminism.* Nerdy But Flirty. https://nerdybutflirty.com/2016/12/02/the-problem-with-the-womens-march-on-washington-and-white-feminism/

Kahn, M. (2017, January 17). *The pussyhat is an imperfect, powerful feminist symbol that thousands will be wearing this weekend.* ELLE. https://www.elle.com/culture/career-politics/news/a42152/pussyhat-project-knit-protest/

Kimball, M. A. (2017). Tactical technical communication. *Technical Communication Quarterly, 26*(1), 1–7. https://doi.org/10.1080/10572252.2017.1259428

Larson, S. (2015, July 23). *This guy 3D prints tombstones and sticks them on packaged meat in the grocery store.* The Daily Dot. https://www.dailydot.com/debug/3d-printed-tombstones-meat-supermarkets/

Manovich, L. (2002). *The language of new media.* MIT Press.

Pine, B. J. (1993). *Mass customization: The new frontier in business competition.* Harvard Business School Press.

Portwood-Stacer, L. (2007). *Do-it-yourself feminism: Feminine individualism and the girlie backlash in the DIY/craftivism movement* [Paper presentation]. Annual Meeting of the International Communication, San Francisco, CA, United States.

https://www.academia.edu/1863481/Do_It_Yourself_Feminism_Feminine_Individualism_and_the_Girlie_Backlash_in_the_CraftivismMovement

Pussyhat Project. (n.d.-a). *Pussyhat project.* https://www.pussyhatproject.com/

Pussyhat Project. (n.d.-b). *Pussyhat project FAQ.* https://www.pussyhatproject.com/faq/

Randall, T., Terwiesch, C., & Ulrich, K. T. (2005). Principles for user design of customized products. *California Management Review, 47*(4), 68–85. https://doi.org/10.2307/41166317

Sherrill, J. T. (2019). *DIY feminism in post-industrial spaces* [Unpublished doctoral dissertation]. Purdue University.

Srnicek, N. (2016). *Platform capitalism* (First edition). Polity.

Walton, R., Moore, K., & Jones, N. (2019). *Technical communication after the social justice turn: Building coalitions for action.* Routledge.

WrrrdNrrrdGrrrl [@WrrrdNrrrdGrrrl]. (2017). *Trans pride flag pussy hat: Because we stand with our sisters, not just our cis-ters.* #WomensMarch #SistersNotCisterspic.twitter.com/ewMsFIO4yA. https://twitter.com/WrrrdNrrrdGrrrl/status/828416317757743109

Chapter Eight

Tactical User Research

How UX Can (Re)Shape Organizations

Guiseppe Getto

User research is now at the heart of many organizations that develop products and services ranging from mobile applications to educational programs. It is at the heart of organizations as disparate as major corporations that provide videoconferencing software and small regional hospitals that want to enable their patients to better cope with illnesses on an outpatient basis. It is a method used by a variety of professionals to understand the needs and values of their primary stakeholders in order to tailor their products to specific types of users and their specific pain points. For the sake of this chapter, we will call these organizations "user-positive organizations." What distinguishes user-positive organizations is that they are careful to advocate for their users so that the problems users face are at least as important as, if not more important than, their own organizational goals. Inquiry into human behavior drives much of what user-positive organizations do. They want to understand their users intimately so that they can better serve them, and they devote significant resources to doing so. At the same time, however, some organizations that carry out extensive user research programs are also very motivated to leverage insights from users to make a profit, so even user-positive organizations are not necessarily motivated purely by user advocacy.

Based on my own experience with both academia and the private sector over the past 10 years or so, which has involved not only hundreds of hours of empirical research with users but also conversations, interviews, and surveys with scores of user experience (UX) professionals, I believe this user-positive approach to be the norm, meaning that it is normal within organizations who know what UX is to claim that they are user positive. From these anecdotal and research experiences, in other words, I have gleaned that it is now the norm in organizations that do UX to claim a user-positive stance, to *claim* that they are advocating for their users and that user research is at the heart of what they do. This has become so much the norm, in fact, that to claim otherwise, to claim that an organization is *ignoring* its users in favor of larger concerns, or even manipulating them in favor of those concerns, has become taboo within the UX community. The community has even come up with its own term to describe using the principles of UX against users: "Dark UX" (Miquel, 2014).

At the same time, I have also encountered people in organizations that are best described as *user theatric*, a term I'm adapting from a recent article by Jesse James Garrett, a pioneer in UX design. A user-theatric organization is one in which the organization creates the "appearance of due diligence and a patina of legitimacy that's just enough to look like a robust design process to uninformed business leaders and hopeful UX recruits alike" (Garrett, 2021). Such an organization is using UX principles without fully committing to them, in other words. Or worse, it is using UX principles in a *dark* manner, to manipulate users into consuming products or services by using the organization's insights against them.

Drawing on recent work at the intersections of UX design and technical communication, in the chapter that follows, I thus sketch out a framework for using UX workflows to reshape organizations in a user-positive manner, a framework I term *tactical user research*. I argue that UX researchers, by virtue of their work to align design workflows with organizational goals and user goals, necessarily work at the crossroads of a variety of important organizational processes, such as cross-functional teamwork, product development, stakeholder engagement, user advocacy, and strategic planning. From this vantage point, if UX researchers adopt a tactical approach to their work, they can radically reshape their organizations to be more ethical by putting user needs at the forefront of organizational goal setting.

To make this argument, I draw on my own experiences as a user researcher, and specifically on my experiences as a user researcher for SeaMe, a grant-funded mobile application for recreational boaters. As I explain below, the small organization behind SeaMe is a user-positive one because we are working hard to balance user needs with organizational goals. In fact, the business model of this budding organization is tied intimately to the needs of users and the government regulators who ensure their safety. Though SeaMe must generate funding in order to exist, the primary goal of the organization is to reduce pain points for users and thus to increase their safety.

A Brief Introduction to SeaMe

Recreational boating currently accounts for a significant portion of recreation-based accidents, fatalities, and property damage. In 2019, for instance, the Coast Guard reported 4,168 accidents that involved 613 deaths, 2,559 injuries, and approximately $55 million of damage to property as a result of recreational boating accidents (American Boating Association, 2021). At the same time, according to industry insiders interviewed as part of this research, but who were guaranteed anonymity by researchers, boating regulations have largely remained unchanged since the 1970s, largely due to successful lobbying by boating manufacturers. From these interviews, I also learned that most state budgets are currently maxed out when it comes to boating safety, meaning there is no funding available to build additional programming to decrease incidents.

Within this context, I was asked to serve as entrepreneurial lead on an NSF I-Corps[1] grant in order to do user interviews regarding the development of a free mobile application that might increase boater safety. The nascent business model of this application involved an exchange of a free mobile application for user data (figure 8.1).

Essentially, government regulators of recreational boating safety, including state representatives known as boating law administrators (BLAs), waste considerable percentages of their budgets every year on expensive search-and-rescue operations to find lost boaters. Recreational boaters, on the other hand, want applications that can help them boat safely while on the water. The SeaMe mobile app would thus deliver deidentified, anonymized data on boater positions if they get lost to emergency responders. In exchange, boaters would receive a free, full-featured safety application that would run on a mobile phone.

This seemingly simple solution, however, has involved considerable research and development and is still not ready for full deployment. At the

Figure 8.1. The SeaMe Business Model. *Source:* Created by the author.

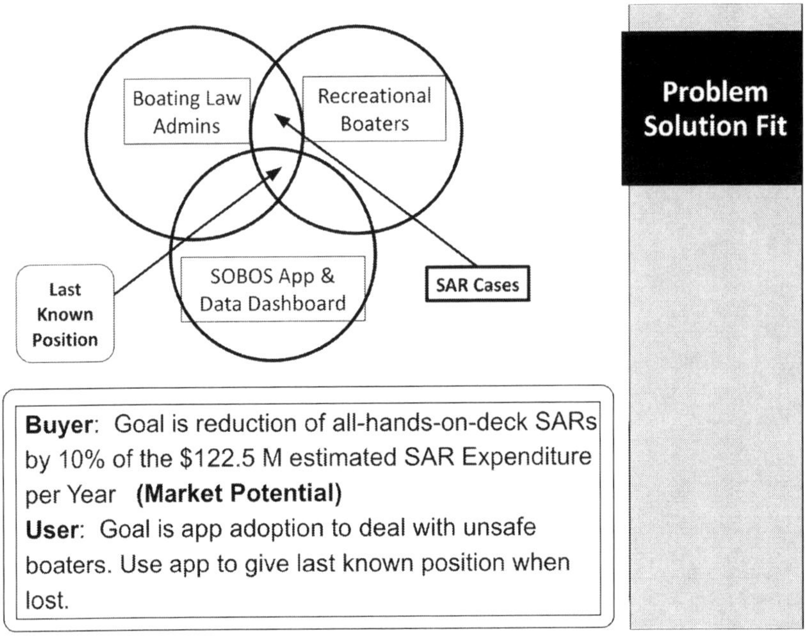

same time, the process of getting to this conceptualized solution has involved a type of user research that is tactical in that it stands to considerably reshape several organizations involved with recreational boating safety. And it is user positive in that it advocates for user needs as a concern equal to other concerns (such as organizational goals). In this chapter, I shall delve more into what I mean by a user-positive framework.

Toward a User-Positive Framework for UX

Recent scholarship at the intersections of UX and technical communication (Albers, 2003; Andrews et al., 2012; Getto, 2020; Getto & Moore, 2017; Getto & St. Amant, 2014; Mara & Mara, 2015; Potts, 2013; Potts & Bartocci, 2009; Redish & Barnum, 2011; Robinson, & Lanius, 2018; Sun, 2013) necessarily supports a tactical approach. I say necessarily because it is indeed the norm in academic research on UX to value users and to advocate for them. At a broader level, this is arguably a founding principle of the academic field of technical communication. Few, if any, members of this field could imagine justifying communication tactics that take advantage of users in favor of organizational goals. In fact, some of us have fled work in the private sector *specifically to avoid* these very tactics. And some of us have fought our own academic organizations when they attempt to deploy tactics that we feel undermine their educational mission.

This is not to create an unnecessary binary between academia and industry. Rather, it is to indicate that, at a fundamental level, all organizations function in a similar fashion when it comes to their users: organizational goals surrounding a given application often shift and change. Target users for the application sometimes change, either because changes to organizational goals require it or because initial testing reveals the application isn't appropriate for its original target user base. Exigences such as these mean that as UX researchers, "we cannot consistently predict what kinds of information might be important to specific groups and in specific situations, we need methods by which we can understand the dynamic relationships between users and technologies" (Potts, 2009, p. 285). In other words: as digital products and services become increasingly pervasive and increasingly complex, the relationships among users, technologies, and contexts of use become increasingly complex and increasingly unpredictable. Even despite this increase in the pervasiveness, complexity, and unpredictability of use cases for digital products and services, or perhaps because of it, "most users are involved in the design process too late to influence the final product" (Andrews et al., 2012, p. 124). This failure to account for users and their contexts "explains systems which function technically but fail because of lack of user acceptance" (Albers, 2003, p. 270).

Instead of viewing existing organizational workflows as rigid and immutable, in other words, tactical user research treats networks of human and nonhuman actors (i.e., people, technologies, modes of communication, supplies, materials, etc.) within organizations as reconfigurable, meaning changeable. We have only to draw on Latour (2005) and his actor-network theory (ANT), which, as Spinuzzi (2008) explains, provides a political and rhetorical view of organizational networks that foregrounds the continual recruiting of new allies—both human and nonhuman—to strengthen particular network configurations over others (p. 25). Latour (2005) encapsulates this process as a shift from "matters of fact" to "matters of concern" (p. 117). Such an approach puts organizational workflows solidly within the realm of rhetoric, defined as a system of communication that relies on probabilistic reasoning.

And if organizational workflows are rhetorical, then that means that they rely on *metis* or a type of cunning knowledge rooted in local expertise (Grabill, 2007, p. 84). For de Certeau (1984), drawing on the work of other scholars, metis is a "form of practice that is always 'immersed in practice,'" thus making it a "form of practical intelligence" that is close to his concept of tactics (pp. 82–83). Of course, for de Certeau, "tactics" (p. 37), or everyday acts of resistance deployed by individuals, are counterposed to "strategies" (p. 36), which are the overall aims of organizations. So, to synthesize this: metis, or local, practical knowledge, is generated through the recurrent activities of individual actors and often takes the form of acts of resistance (tactics). Strategies deployed by organizations often fail because the actors they target resist them.

From a UX perspective, these recurrent activities of organizational actors can help dominate users and encourage them to buy products and services that are not useful to their specific lifeways. When this happens, users often resist these activities by utilizing technologies in ways unintended by designers. When users do this, they often draw on their local knowledge, or metis, which is "subversive, a way of inventing knowledge and persuasive discourse that seeks to counter domination" (Grabill, 2007, p. 85). From this context, tactical user research is not only an approach but is also "a type of conduct" for technical communicators (Miller, 1989, p. 23). Tactical user research invites technical communicators to not only work proficiently to help sustain organizational networks but to act for "the good of a community" by actively caring for the actors involved in organizational networks and their knowledge-making processes (Miller, 1989, p. 23). Following Hart-Davidson and colleagues (2008), such a shift "suggests a role for the technical communicator as advisor or consultant" (p. 10).

The tactical-user-research framework thus helps researchers, practitioners, and teachers to accomplish the following:

- Identify key actors within organizational workflows
- Understand how these actors stabilize existing network configurations
- Use this knowledge to create opportunities to destabilize existing configurations in favor of new ones

For the purposes of this chapter, I will be focusing on how I used tactical user research to accomplish these goals, because I want to emphasize how important this framework is to the overall UX process and how neglecting this situated approach can lead to a user-theatric or dark-UX approach.

But first, let me define the important method behind tactical user research. User research can be defined as the process of interviewing users of an application (or potential application) about the context in which they will use the application. Though it can be deployed at nearly any stage of the design process, user research, if it is employed at all, is typically[2] applied in the following fashion (Buley, 2013; Morville, 2007; Garrett, 2003; Hoober, 2014; Hartson & Pyla, 2012):

1. Preliminary research: user interviews within target use context
2. Prototyping
3. Usability testing of prototype
4. Maintenance

Essentially, design projects often start with user interviews, preferably conducted in the context in which they will be using the application. These interviews might be followed up with observational sessions with users in which UX researchers note common work practices, technology usage, and other elements of the users' context. From this contextual data, a rough prototype of the application is developed. In the past, this has commonly started with the development of a paper prototype, which is still the case according to my anecdotal interactions with practitioners, but often quickly proceeds to the development of a low-fidelity, clickable prototype that can be used in usability testing.[3] This prototype is then refined through succeeding rounds of usability testing until it reaches high fidelity and then is finally launched as a product or service. Maintenance of the product or service often entails updates, design tweaks, and content strategy for the product, with the design process beginning in earnest again when an exigence for major changes arising, such as major changes to web technologies or organizational goals.

Below I describe the process of tactical user research, my own specific approach to this method, and how it has led to a more tactical approach to the development of a mobile app for recreational boaters.

A Mini Case Study in Tactical User Research

For my user research on the SeaMe app and following the list given in the previous section, the first step was thus to identify key actors within the organizational workflows surrounding recreational boating safety. The important word here being *key*. To date, I have collected interviews from 141 stakeholders during the initial NSF grant devoted to this purpose. I encountered many different actors during these interviews, including recreational boaters, federal regulators, state regulators, boat manufacturers, and many others.

By saying the first step of tactical user research is to identify key actors in organizational workflows, however, I mean it's important to differentiate from the diverse network of people and technologies that make up any given community who the most important or most influential actors are.

Step 1: Identify Key Actors Within Organizational Workflows

One of the first challenges of tactical user research is identifying key actors within organizational workflows. In UX theatric workflows, this has often been oversimplified under the rubric of *users*. This approach holds that the key actors that UX should focus on are users. The problem for someone like Latour (2005), however, is that *users* is far from descriptive enough to identify the impacts of all the objects involved with a specific network. As I began interviewing stakeholders for a nascent mobile boating-safety app in collaboration with the app's development team, for example, I found a variety of humans and nonhumans that were somehow involved in the network we might term *recreational boating safety in the United States*.

I found that these actors play various roles in this network and help establish the matters of concern central to it.

With such a panoply of actors present, identifying all of their roles within such a network can indeed be challenging. This is where a method such as tactical user research really shines, however. Eliciting stories from stakeholders using sound user-interview practices can help flesh out a network configuration. Two such stakeholders are particularly important in such an endeavor:

- Intended users for the application
- Subject-matter experts (SMEs) responsible for maintaining existing network configurations

While user-theatric organizations focus solely on the former (users), neglecting SMEs in positions of authority within the network in which an application will be launched can lead to disempowering the users we're supposed to be serving as UX professionals. This is because there are always barriers to the development of new applications.

In the SeaMe mobile-app network, for instance, two main actors would become key to the development of the app:

- Recreational boaters piloting craft that are smaller than 24 feet
- State BLAs

As can be seen from figure 8.1, the main crux of the SeaMe app is that BLAs have limited budgets for search-and-rescue (SAR) operations to find lost boaters. At the same time, my research revealed that recreational boaters piloting craft smaller than 24 feet often lack the necessary technology to signal to first responders (who are employees of BLAs) that they are lost. In this case, the boater's last known position is unknown to first responders, and the first responders have to search extensively for them. Because these first responders are operating on a limited budget, these searches are measured by how much a given state can afford to spend per boater.

I would not have discovered this key connection between user needs and organizational needs, however, had I not interviewed a wide variety of SMEs in boater safety. This is why, in order to identify the key actors in a given network, I actually recommend starting by interviewing SMEs, not users. This may seem to fly in the face of a user-positive framework, but the point of starting with SMEs is to identify current limitations and opportunities within a given network. In order to frame an application as a positive value for users, the first step is to assess its feasibility, in other words. If an application can't be funded, launched, and maintained by an existing network, then it will never reach users. Also, talking to SMEs can help pinpoint the types of users they currently serve. When talking to BLAs, for instance, we learned that for every state described in the interviews, finding lost boaters was the main pain point.

As far as how to find potential actors, there are a variety of approaches, including the following:

- Approach leaders of existing organizations within a network to ask for their role in helping users. In my own experience, we approached boating membership clubs, governing bodies, and state regulatory agencies.

- Use social media tools such as LinkedIn Sales Navigator (https://business.linkedin.com/sales-solutions/) to identify and message potential stakeholders to ask for their opinions.
- Use the snowball method: ask every interviewee for two or three additional people like them that might be willing to be interviewed.
- Keep a detailed database of all potential interviewees is crucial as one begins to flesh out a given network and its actors.

Concerning the interviews themselves, Thornton (2019) has developed a very comprehensive set of guidelines for conducting and analyzing user interviews (guidelines that can be used for any stakeholder interview). They include the following:

- Don't ask leading or directed questions.
- Don't ask people what they want.
- Ask open-ended questions.
- Don't ask yes/no questions.
- Don't make assumptions. Ask the stupid questions.
- Have a set of questions you use every time.
- Ask the same question from multiple angles.
- Never mention other users.
- Ask follow-up questions.
- Synthesize findings and make recommendations.

Below are my own interview scripts for the key actors I've identified, recreational boaters and BLAs:

Interview Questions for Recreational Boaters

1. How long have you been boating?
2. How do you communicate with people onshore while you are boating?

3. What are challenges when you're trying to communicate with people onshore?
4. How do you deal with those challenges?
5. How do you deal with conditions on the water (i.e., weather, hazards, other boaters, etc.)?
6. What are challenges you experience with conditions on the water (i.e., weather, hazards, other boaters, etc.)?
7. How do you deal with those challenges?
8. Is there anything I missed?
9. Do you know two or three more people we could talk to?

Interview Questions for BLAs

1. Tell us about what you do for recreational boaters within your current role.
2. What are your biggest frustrations do you have within your current role when serving recreational boaters?
3. How do you currently deal with those frustrations?
4. What would make your job easier to do?
5. What kinds of data do you use to make decisions in your job?
6. Where do you currently get that data?
7. What types of data do you not have that you need?
8. How would you need that data delivered to you (i.e., in a technology, in raw form, etc.)?
9. Anything I missed that you think is important?
10. Do you know two or three additional people we could talk to?

It's important to note that these scripts were deployed in a semistructured manner. Tactical user research is as much an art as a science. Asking open-ended questions, digging for assumptions, and asking follow-up questions, as Thornton (2019) advocates, are all important ways to try to get at the foundational beliefs of actors.

Identifying nonhuman actors and their roles in a network, on the other hand, is synonymous with what UX designers term *requirements gathering*: the process of understanding the core technologies and technical systems that will support a new application. Most developers, and even UX designers, are unfamiliar with a framework such as actor-network theory, so requirements gathering is the right terminology to use with nonacademic stakeholders. Regardless, it's essential to understand how the core technologies and technical systems within an application's network:

- Require, encourage, or constrain certain types of human activity
- Are accessible or inaccessible to certain types of humans in the network
- Match, or fail to match, the expectations of humans within the network
- Interact with other nonhuman actors, such as existing applications already being used by humans

Seeing nonhuman actors as stakeholders that can require, encourage, or constrain certain types of human activity is a key differentiator here. In the example of the SeaMe mobile app, the development team chose to focus on asking users for deidentified data in order to protect their right to privacy. This required assembling data-gathering systems that scrub all data of personal identifiers. Had we decided to ask for private user data, this would have required a completely different approach to the development of the app, an approach bordering on dark UX approaches applied by companies such as Facebook that extract personal data from users in order to sell it to advertisers, thus providing a "free" service that actually has a very high social cost.

Regardless, once key actors are identified, it's time to understand how these actors stabilize existing network configurations. Of course, all the stages of tactical user research are less linear steps than heuristics that are applied recursively. It's very common to interview several stakeholders only to discover that one is on the wrong path, as I did when interviewing members of marine-insurance companies who wanted types of data we couldn't easily provide. It's also common to develop a new understanding of a network configuration that requires validation with an entire new group of interviews or a return to previous groups. In any event, I discuss the process of ascertaining network configurations in the next section.

Step 2: Understand How These Actors Stabilize Existing Network Configurations

Once key actors are identified within a given network, the next stage of tactical user research involves ascertaining how these actors stabilize existing network configurations. In Latourian parlance, and following Spinuzzi (2008), a particular network configuration becomes stabilized through the continual recruiting of new allies—both human and nonhuman—to strengthen particular connections within the network (p. 25). Latour (2005) encapsulates this process as a shift from "matters of fact" to "matters of concern" (p. 117). Essentially, what stakeholders within a network consider to be factual elements of the network are actually just elements that have been strengthened through continual activity. Such an approach puts organizational workflows solidly within the realm of rhetoric, defined as a system of communication that relies on probabilistic reasoning. In a real sense, then, understanding existing network configurations is essential to ascertaining the available means of persuasion in the Aristotelian sense when dealing with complex networks that involve both humans and nonhumans.

Detecting these configurations involves what Potts (2009), again following Latour (2005), has characterized as the act of tracing connections between actors. Elsewhere, I have characterized this method as examining a network for rhetorical impacts between actors or "building a rhetorical awareness of the impacts people, technologies, and other resources are having on one another" (Getto et al., 2018, p. 182). Two specific types of impact that interest me when developing a new application involve Latour's view of how meaning is developed within a network. An intermediary, Latour (2005) writes, "transports meaning or force without transformation" (p. 39). Mediators, unlike intermediaries, "transform, translate, distort, and modify the meaning or the elements they are supposed to carry" (p. 39). In other words, actors who describe mediators transform meaning, whereas actors that describe intermediaries do not transform meaning.

This simple classification of intermediary versus mediator can help a researcher understand which actors within a network are having impacts, meaning modifying meaning, and which are simply serving as channels for existing meaning. To help exemplify this difference, consider figure 8.2, an NSF I-Corps model for understanding how a startup business can enter into an existing marketplace through creative disruption.[4]

You will see in this figure a depiction of variety of actors and their impacts on the SeaMe business model. Table 8.1 categorizes these actors as mediators or intermediaries.

What is important to take from this ecosystem is that it depicts both actors that will alter meanings that support network existing configurations (mediators), thus potentially encouraging new network configurations, and actors that will maintain status quo (intermediaries), thus supporting existing configurations.

In practical terms, discerning intermediaries and mediators within a network involves synthesis and analysis of interview data. A researcher should code or otherwise look for patterns among interviews in order to understand how a given human actor impacts, or fails to impact, the network. Are they simply serving as an intermediary, in other words? Or are they serving as a mediator by modifying meaning through their actions? Classifying actors in

Figure 8.2. The SeaMe Business Ecosystem Map. *Source:* Created by the author.

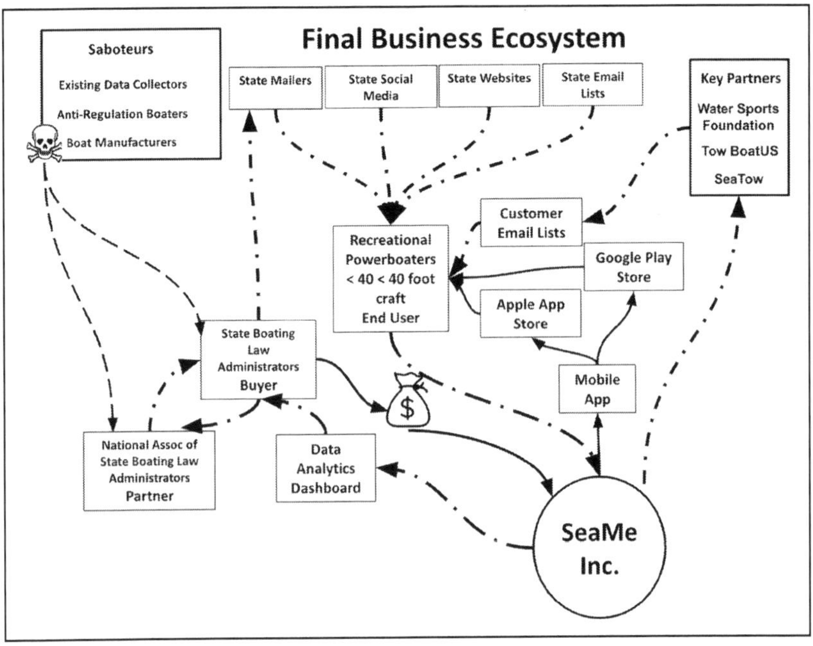

Table 8.1. Key Actors as Mediators and Intermediaries

Actor	Role in Network	Mediator or Intermediary?
State Regulator	Payer: Purchases data from the SeaMe mobile app in order to reduce all hands-on-deck search-and-rescue missions, thus saving money that can be deployed for prevention	Mediator
Boat Manufacturer	Saboteur: Doesn't want new data deployed that might disrupt status-quo marketplace in which regulation is low and resulting production costs are low	Intermediary
Boat Insurer	Influencer: Promotes the SeaMe app to customers in exchange for access to data that will help them reach lost boaters faster	Mediator
Boating-Safety Educator	Influencer: Promotes the SeaMe app to customers in exchange for access to data that will help them provide better safety information to boaters	Mediator
Data Collectors	Saboteur: Doesn't want new data-collection methods deployed that might disrupt status quo of them doing costly surveys	Intermediary
Recreational Powerboater operating < 40-foot craft	Data Generator: Provides last known position to state regulators when lost in exchange for a full-featured, but free, mobile app	Mediator
Recreational Boater who is against any type of regulation	Saboteur: Refuses to adopt mobile safety app for fear of data (even deidentified) being shared with outside parties	Intermediary
Mobile App Marketplaces	Channel: Provide mobile phone users with access to downloadable apps native to their respective platforms (i.e., Android or iPhone)	Intermediary
Data Analytics Dashboards	Data Purveyor: Provide state regulators with customized reports based on aggregate data	Mediator
State-Level Email Lists of Recreational Boaters	Channel: Provide a means of reaching individual boaters who are registered with a given state in order to promote SeaMe app	Intermediary

Source: Created by the author.

this way allows the tactical user researcher to better ascertain the available means of persuasion within a given network or to create opportunities to destabilize existing configurations in favor of new ones, which I turn to next.

Step 3: Use This Knowledge to Create Opportunities to Destabilize Existing Configurations in Favor of New Ones

The question now becomes: How did I go from figure 8.2 to figure 8.1 as a user researcher? Or, in other words: How did I identify the key actors involved with the SeaMe app as (1) recreational boaters operating craft smaller than 24 feet and (2) BLAs? How did I discover that the key to the development of the SeaMe app is that recreational boaters want a mobile app that will help them remain safe while on the water, and BLAs want boaters' last known position so they can rescue them if they get lost? The answers to these questions are revealed by thinking about how the network of boater safety is *currently configured* and how it *could be configured.*

The key lynchpin of the SeaMe mobile app business model is that it would provide a boost to recreational safety by allowing boaters in distress to transmit their last known location to BLAs when they are lost (again, see figure 8.1). In exchange, BLAs will save money on not having to deploy costly search-and-rescue missions that cover large areas. This will leave additional money available to deploy for safety programs that *prevent*, rather than react to, boating incidents. When looking at the full ecosystem of this marketplace, however, it is clear that there are several actors currently working to maintain the existing configuration because it is beneficial to them. Through interviews with several BLAs and other types of regulators, for instance, we learned that boat manufacturers have actively lobbied politicians against passing life-saving policies, such as a mandatory national life-jacket law.

This was explained to us by participants, both those within boat-manufacturing organizations and those serving as state or federal regulators, as a means of keeping profits high. When consumers view recreational boating craft as safe, fun vehicles, they are more likely to spend on them. If they perceive them as they actually function within the recreational boating network—as both fun vehicles and vehicles that can lead to serious injury, property damage, or even death—then consumer confidence drops, and sales drop. In a more generous view of this activity, one participant who serves as the head of safety for a large boating manufacturer felt that existing regulations were sufficient to prevent as many incidents as could be prevented. He did not feel that there was any way to further reduce boating incidents, which he chalked up to "operator error." From his view of the recreational boating network, in other words, he was acting in a just manner toward consumers, and further regulation would simply harm his organization's ability to stay in business while not benefiting consumers.

Destabilizing existing network configurations in favor of new ones, in other words, is not easy, especially when existing network configurations are densely woven, based on years, or decades, of recurrent activity. Boating regulations remain largely unchanged since the 1970s, even though the number of boaters has greatly increased since that time. Actors who serve as intermediaries by transporting existing meanings, essentially dominating other actors through their recurrent activity, serve to squelch alternative meanings. There is a reason that state regulators tap out their resources every year rescuing lost boaters rather than engaging in prevention. And that reason is found in the recurrent activities of actors that maintain that status quo due to their own interests.

At the same time, had I found no possibility for an alternative network configuration, no user-positive framework by which to proceed, this would be a much darker story. It is arguable that boat manufacturers are using a dark UX approach to the construction of their vehicles, for instance, by chalking up all existing boating incidents to operator error that can never be altered. As one participant, a boating educator, put it, anyone with sufficient funding can go decide to buy a boat with no training, no safety equipment, and no safety expertise and be out on the water that afternoon, completely unaware of the dangers. I also detected some user-theatric activity, such as one participant heavily involved with boat manufacturers who confessed to us that most recreational boaters are "stupid," citing prop-strike incidents that happen every year despite him being "forced" to put a notice on new boats warning of this possibility. This sort of defense of the status quo that users are simply misusing a technology while maintaining public appearances of improving usability is the hallmark of a user-theatric approach.

In light of all this, the SeaMe mobile app represents a user-positive alternative to these existing configurations and thus is a tactical approach to the overall network of recreational boating safety. If my efforts as a consultant can help the SeaMe development team deploy a free mobile app that is adopted by a broad swath of recreational boaters, the following configurations may be shifted:

- Boaters will receive free safety information, including hazard detection, accurate charts, and weather warnings
- State regulators will receive the last-known position of boaters who are lost or in distress, thus lowering the amount of costly search and rescue missions they must deploy each year
- State regulators will be able to use money previously spent exclusively on search and rescue missions to create more preventative programs that are currently beyond their budgetary reach
- Because of all this, boating incidents may decrease
- Such a decrease may persuade existing intermediaries to support this new network configuration over the old one

As with all rhetorical work, this effort is probabilistic, meaning possible but by no means guaranteed. It remains to be seen if we can develop and deploy a mobile app that serves recreational boaters and state regulators while avoiding, or altering, the actions of powerful actors such as boat manufacturers and existing data collectors. Overall, our efforts are not to pit such actors as the villains in this network but rather to persuade them that there is a more just configuration of this network that treats their users in a more positive light. And, of course, there are many allies within the network whose missions align with our own, such as boat insurers and boat educators whose organizational goals center on improving boating safety. I shall close with some limitations for this approach and some opportunities for future work in this area.

Conclusions, Limitations, and Opportunities for Further Work

The tactical user research framework is not a revolutionary break from UX best practices, but rather it is a *return to them*. Though I am solely an academic researcher of these trends, and a sometime UX practitioner, I heartily agree with Garrett's concerns over user-theatric approaches. There was a time, in fact, when I considered leaving academia for the world of industry to be a full-time UX practitioner. What stopped me was getting to know my new potential colleagues in that space and realizing, to my shock and dismay, that many of them did not view UX as a means to actually improve the experiences human beings have with technology but rather as a means to impress corporate executives and boost their careers. This is certainly not the case with thought leaders and user advocates such as Garrett and many of the other industry practitioners cited in this chapter, but it is very common among rank-and-file UX practitioners I have personally encountered.

These user-positive practitioners would argue that the purpose of UX should be to reshape organizations. There are limitations, however. There are reasons why a user-theatric approach has taken hold. And the reason is the status-quo mentality of existing organizations that deploy UX activity. As long as it is easier to dupe users or to use them as matters of concern for existing organizational goals, then there will be UX practitioners who deploy its methods for their own personal gain. There is perhaps no industry immune to self-centered professionals. And from the outside, it can be difficult to discern whether an organization is truly user positive or just playing this role.

This limitation should not discourage user-positive researchers and practitioners from pursuing projects that can have real impacts on the lives of the human actors they serve, however. In the example of my own project, it is far from guaranteed that a free mobile app will disrupt the existing configuration of the recreational boating network, but this is the work of tactical user research. If we are to treat users as human being worthy of respect and strive to continually deliver better experiences to them, then we will have to become savvy about existing networks that these users inhabit. We can't afford to assume that simply deploying UX best practices are enough to empower users. We must also use our knowledge of rhetoric to persuade actors within networks to empower users.

Acknowledgments

This work was partially supported by an NSF I-Corps Grant.

Notes

1. The NSF I-Corps program has its own trademarked approach to business development that is based largely around the work of Steve Blank (https://steveblank.com/).

2. By "typically," I mean to indicate best practices articulated largely from practitioners in the field. There are no known empirical studies by academic researchers that systematically examine the UX process and how it is deployed.

3. These interactions included several workshops with some of the top UX researchers in the field, including Leah Buley, James Kalbach, and Alberta Soranzo, all of whom described the utility of beginning design projects with paper prototypes.

4. SeaMe was originally called SOBOS, which stood for the Self-reported On-the-water Boat Operator Survey.

References

Albers, M. (2003). Multidimensional audience analysis for dynamic information. *Journal of Technical Writing and Communication, 33*(3), 263–279. https://doi.org/10.2190/6KJN-95QV-JMD3-E5EE

American Boating Association. (2021, June 10). Boating fatality facts. https://americanboating.org/boating_fatality.asp

Andrews, C., Pohland, E., Burleson, D., Unks, S., Scharer, J., Elmore, K., Wery, R., Lambert, C., Wesley, M., Oppegaard, B, & Zobel, G. (2012). A new method in user-centered design: Collaborative prototype design process (CPDP). *Journal of Technical Writing and Communication, 42*(2), 123–142. https://doi.org/10.2190/TW.42.2.c

Buley, L. (2013). *The user experience team of one: A research and design survival guide.* Rosenfeld Media.

de Certeau, M. (1984). *The practice of everyday life.* University of California Press.

Garrett, J. (2003). *The elements of user experience: User-centered design for the web.* New Riders.

Garrett, J. (2021, June 3). *I helped pioneer UX design. What I see today disturbs me.* Fast Company. https://www.fastcompany.com/90642462/i-helped-pioneer-ux-design-what-i-see-today-horrifies-me

Getto, G. (2020). The story/test/story method: A combined approach to usability testing and contextual inquiry. *Computers and Composition, 55*, 1–13. https://doi.org/10.1016/j.compcom.2020.102548

Getto, G., & Moore, C. (2017). Mapping personas: Designing UX relationships for an online coastal atlas. *Computers and Composition, 43*, 15–34. https://doi.org/10.1016/j.compcom.2016.11.008

Getto, G. & St. Amant, K. (2014). Designing globally, working locally: Using personas to develop online communication products for international users. *Communication Design Quarterly, 3*(1), 24–46. https://doi.org/10.1145/2721882.2721886

Getto, G., Franklin, N., Ruszkiewicz, S., & Labriola, J. (2018). User experience in a networked environment: How Latour can help us do better UX work. In K. Moore & D. Richards (Eds.), *Posthuman praxis in technical communication* (pp. 176–196). Routledge.

Grabill, J. (2007). *Writing community change: Designing technologies for citizen action.* Hampton.

Hart-Davidson, W., Bernhardt, G., McLeod, M., Rife, M., and Grabill, J. (2008). Coming to content management: Inventing infrastructure for organizational knowledge work. *Technical Communication Quarterly, 17*(1), 10–34. https://doi.org/10.1080/10572250701588608

Hartson, R., & Pyla, P. (2012). *The UX book: Process and guidelines for ensuring a quality user experience.* Morgan Kaufmann.

Hoober, S. (2014, May 5). *The role of user experience in the product development process.* UX Matters. http://www.uxmatters.com/mt/archives/2014/05/the-role-of-user-experience-in-the-product-development-process.php

Latour, B. (2005). *Reassembling the social: An introduction to actor-network-theory.* Oxford University Press.

Mara, A., & Mara, M. (2015). Capturing social value in UX projects. *Proceedings of the 33rd ACM International Conference on Design of Communication, 23,* 1–6. https://doi.org/10.1145/2775441.2775479

Miller, C. (1989). What's practical about technical writing? In B. E. Fearing & W. K. Sparrow (Eds.), *Technical writing: Theory and practice* (pp. 14–24). MLA.

Miquel, M. (2014, July 7). *Throwing light on dark UX with design awareness.* UX Magazine. https://uxmag.com/articles/throwing-light-on-dark-ux-with-design-awareness

Morville, P. (2007, July 23). *User experience strategy.* Semantic Studios. http://semanticstudios.com/user_experience_strategy/

Potts, L. (2009). Using actor network theory to trace and improve multimodal communication design. *Technical Communication Quarterly, 18*(3), 281–301. https://doi.org/10.1080/10572250902941812

Potts, L. (2013). *Social media in disaster response: How experienced architects can design for participation.* Routledge.

Potts, L., & Bartocci, G. (2009). <Methods> experience design </methods>. Proceedings of the *27th ACM International Conference on Design of Communication, 27,* 17–21. https://doi.org/10.1145/1621995.1621999

Redish, J., & Barnum, C. (2011). Overlap, influence, intertwining: The interplay of UX and technical communication. *Journal of Usability Studies, 6*(3), 90–101.

Robinson, J., & Lanius, C. (2018). A geographic and disciplinary examination of UX empirical research since 2000. *Proceedings of the 35th ACM International Conference on Design of Communication, 8,* 1–9. https://doi.org/10.1145/3233756.3233930

Spinuzzi, C. (2008). *Network: Theorizing knowledge work in telecommunications.* Cambridge University Press.

Sun, H. (2013). *Cross-cultural technology design: Creating culture-sensitive technology for local users.* Oxford University Press.

Thornton. (2019, March 2). *How to conduct user interviews.* UX Collective. https://uxdesign.cc/how-to-conduct-user-interviews-fe4b8c34b0b7

Chapter Nine

The Motivations of the Marginalized

Identifying and Navigating Hegemonic Factors in Petroleum Risk Communication

Joseph E. Williams

> While retiring offshore work crews remove their orange uniforms, they banter in Malayalam, Nepali, Hindi, and Tagalog as they depressurize in the locker room to prepare for time off, occasionally smoking a cigarette in the makeshift plastic-encased smoking-area pod. Supervisors in construction hats carry clipboards and jockey for a moment of the oil installation manager's time; his office, the only one with windows, always appears to be filled with other supervisors, all of whom hail from European countries. Inside the platform are doorways with raised entryways, tight spaces, and a stark absence of greenery everywhere. A faded and warped poster of a tropical beach scene at the stairwell adds a touch of sadness rather than inspiration. The canteen houses cliques of mostly Danish, Filipino, Indian, and Nepali workers. An occasional solitary worker shyly asks where to go or what to do next.

This excerpt from my own field notes points to an environment and constellation of issues and ideas that still require exploration. In this study of risk communication on an offshore petroleum rig, one of the lingering issues I came to see as unsolved and understudied is the role of intercultural communication, excluded ethnic identities, and the effect of an entrenched management and hierarchy that often ignore, implicitly or explicitly, the needs and ideas of these different populations.

So those subordinates, namely foremen and crew, who are all from developing countries must find ways to cope offshore. In the face of this challenge, I advance that a version of tactical technical communication (TTC) can be reframed, with specific reference to language and culture, to study and improve the conditions of workers on these rigs, even as they themselves push back in ways that are tactical rather than strategic, informal rather than formal. TTC offers ways of communicating outside of official or institutional contexts, ways of doing that often work against the grain. The work of Miles A. Kimball has been essential in fleshing out the contours of a communicative praxis that was not meant to be recorded, noticed, or centralized necessarily borne out of the distinction between strategies and tactics introduced in Michael de Certeau (1984)'s *The Practice of Everyday Life*. In discussing the practice of everyday life, de Certeau explores how the invisible remain othered even as they assimilate, saying, "They metaphorized the dominant order: they made it function in another register. They remained other within the system which they assimilated and which assimilated them externally. They diverted it without leaving it. Procedures of consumption maintained their difference in the very space that the occupier was organizing" (p. 215).

Harkening to de Certeau's chapter "Making Do" (2004), foremen and crew strategize as they navigate the tricky waters of life offshore. This navigation is easier said than done: Many subordinates speak very little English, and many offshore supervisors eschew cultural mores of their foremen and crew to the point where a formidable amount of bullying takes place on the oil platform, a dynamic that Nielsen and colleagues (2013, 2014) discuss as a major concern on offshore oil rigs.

In the light of these institutional and hegemonic communicative practices, and the all-too frequent bullying, we can see how other tactics are used by these nonmanagement populations on the offshore platform. Indeed, TTC plays a crucial role in the lives of offshore foremen and crew. There are already significant barriers to communication offshore, including the role of hegemony, the persistence of cultural intolerance, and a range of English-language skills, from being too proficient. TTC allows the bilingual foreman to relay pertinent information to colleagues in a shared tongue so that these workers can carry out their daily job tasks effectively as well as navigate the tricky world of offshore life. These marginalized or otherwise excluded workers need to find their footing within their organization and job descriptions, as well as pivot around so many daily dangerous working conditions. Stepping into the situation is the foreman with power, namely, the power of language: "[W]hether the noninstitutional communicator is in the role of a subject matter expert, she is in a position of power created by the imbalance of knowledge. Even if the communicator is only expressing 'how she did it,' she is participating in a power imbalance" (Randall, 2022, p. 7). Nevertheless, these purported power imbalances create hope and understanding among makeshift translation teams found throughout the offshore worlds of foremen and crew.

One reason these environments have been understudied with respect to their communicative strategies is their obvious isolation, miles away from the coast, reachable usually by helicopter. I, however, had the opportunity to not only visit an Arabian Gulf oil platform of a petroleum company with a global presence but also attend training sessions mandated for all of those employed by petroleum entities. My observations, conveyed here via thick description, show why TTC is key in navigating offshore life for marginalized workers.

Background

Miscommunication and lack of preparedness can be lethal combinations for those involved in the petroleum industry. Effective communication among supervisors and crew and preparedness for risks involved with offshore life can alleviate these risks, but not without a certain amount of buy-in from the stakeholders involved. Under the best of conditions, the effectiveness of this buy-in leads to concepts of mutual respect, teamwork, the role of language, and cultural sensitivity, especially among multiethnic workforces. Discovery Oil (DO), a pseudonym for the actual company name, is a global entity in terms of gas and oil extraction. The company's reputation for safety has led to (re-)training sessions, which all offshore and onshore petroleum employees must attend annually. Unfortunately, these (re-)training sessions do not always promote mutual respect, teamwork, the role of language, and cultural sensitivity, hence the further reliance on TTC among offshore subordinates, most of whom hail from developing countries. In the face of a power structure that is often foreign, these workers often cobble together ways of communicating that echo some of these strategies but that remain the tactics of the workers.

After spending time not only attending (re-)training sessions and spending nights offshore, I discovered a lack of safety-document readability for the intended multicultural audience; lack of English proficiency among offshore subordinates; evidence of makeshift translation teams among offshore subordinates; lack of supervisor consistency in addressing language issues; lack of consistency in supervisor approachability; and evidence of confusion with DO's safety-initiative message of intervention. Hundreds of workers readily accept these conditions in order to bring money back to their respective families. Although there is little research about the lives of foremen and crew, recent TTC scholarly work has been published about the transgender community, another marginalized group in dire need of TTC for their own needs. Edenfield and colleagues (2019) discuss inherent barriers for transgender people to access affordable health care; due to the complexities involved, online environments have witnessed a rise in user-generated instruction sets specifically addressing and providing direction on the self-administration of hormone therapy. One is written, while the other is verbal; otherwise, there are clear parallels between these two marginalized groups. Additionally, Nielsen (2016) discusses the interrelation between places and working-life opportunities, "showcasing how certain types of work and certain types of places can give birth to unconventional working life patterns" (p. 539). These articles show how the lack of inalienable rights and unconventional working conditions reinforce the need of TTC.

Doing Versus Becoming

One signal difference between those versions of technical communication that structure the activities of institutions and organizations and those practiced by individuals appears in the distinction between becoming and doing. As Kimball (2017) discussed, technical communication does not have to be positioned toward experts or the making of experts (becoming); rather, the aim for many consumers is just to get a specific job done. Central to this distinction is Kimball's view that the "organizational assumption obscures a larger view of the technical communication performed by millions of people each day on their own, working outside of, between, and even counter to organizations" (p. 1). Kimball acknowledges the ubiquity of these tactics and the existence of these practices whether they're named as TTC or not.

But this theorization of a common practice has consequences wider than it might seem. Becoming an expert or doing a job has special considerations when technical communication is used in spaces where cultural norms are not inherently respected.

TTC becomes important for international environments and in organizations where a dominant culture and language and other cultures and languages interact. TTC then assists in bridging those gaps between the communication of the institution or organization and those of the employees, workers, or users, not all of whom might be comfortable becoming part of a larger hegemonic system.

Offshore, Tactically Speaking

In my study of technical communication on an offshore platform, one thing that seemed clear to me was that I had reached a kind of terra incognita, a place where national identity was seemingly muted, and a place that previously had not been the subject of study with respect to technical communication. But as I have revised and continued this work, I have been intrigued by the difference in making do among these international populations in an international location.

According to DO, oil-rig subordinates are "frontline guys from 'yessir' cultures." In other words, these workers are reticent to admit that information is not understood properly, information such as company mandates or instructions. Rather than admit confusion with meaning of message, cultural mores encourage group harmony; hence, some employees do not say anything at all. Such behavior is indicative of collectivist cultures of subordinate inhabitants of DO rigs and platforms. Hofstede and colleagues (2010) discuss attributes of collectivist cultures, one of which is group harmony. What is easy for individualists commonly found in the Global North to do, such as requesting clarification, can be perceived as shaming or questioning the honor of the group for collectivists: "[T]he virtues of harmony and maintaining face reign supreme. Confrontations and conflicts should be avoided or at least should be formulated so as not to hurt anyone" (Hofstede et al., 2010, p. 118). To complicate matters further, DO says that there are multiple (approximately three) levels of English language skills; the lower down the chain of command, the lesser the amount of English is spoken typically.

Offshore volatility contributes further to the hazards of life on the oil platform. Bullying can also be a problem offshore, especially between various factions and cliques. Nielsen and colleagues (2013) cite various factors that can complicate a sense of well-being offshore: "Working on offshore installations involves a range of risks such as accidents, fires, explosions, and blowouts. Another feature is the tight social interactions during the shifts on the offshore installations. Severe interpersonal conflicts may complicate communication and collaboration between personnel in safety critical operations" (p. 368). The constant awareness of multiple environmental and societal dangers, which may be further complicated by intercultural miscommunication or linguistic issues, provides a veritable hotbed of challenges for intercultural workers to overcome on an oil platform. And then disaster strikes.

A Quick Portrait of an Oil-Rig Disaster

Deepwater Horizon is a quintessential example of offshore oil drilling gone wrong. Faulty equipment and a neglect of safety checks and balances within British Petroleum's (BP) administration created a cataclysmic chain reaction that resulted in months of leakage and insurmountable decimation of oceanic ecosystems in the Gulf of Mexico. In the *Deepwater Horizon* catastrophic event, an oil / gas fluid explosion, so swift that major failsafe safety operations were never initiated by its offshore employees, blew out with a force that killed 11 crewmen from flames that could be seen 40 miles away (Crittenden, 2010). While *Deepwater Horizon* and its aftermath are an extreme case, a multitude of examples point to the necessity to further improve and enhance risk communication on oil rigs and platforms (Wandekoken et al., 2011; Skogdalen et al., 2011; Skogdalen & Vinnem, 2011; Skogdalen & Vinnem, 2012; Lundberg et al., 2009; Leveson, 2004).

In order to prevent another event such as what occurred on the *Deepwater Horizon*, safety training is paramount onboard oil rigs and platforms, places in which there are chemical and mechanical risks that must be addressed on a daily basis. To that end, DO's safety initiative encompasses different kinds of company media (annual safety [re-]training sessions, looping lobby videos, and newsletters, to name a few) in order to drive home this message of safety. But these institutional initiatives can only reach so far, as the existence of risky behaviors on the rig makes clear. According to DO, daily risk is further compounded by subordinates' high-risk and purportedly illegal job practices such as "hot work," which involves the makeshift use of a flame to cut corners and mend or weld metal on oil rigs and platforms. DO has noted that the unauthorized use of hot work still exists in plentiful amounts aboard their oil rigs and platforms and is also an affront to the company's safety initiative. DO is specifically concerned with hot work as an example of risky behavior, and it is attempting to abolish all unauthorized hot work since high temperatures could cause gases to combust and lead to a full-scale rig explosion.

Intercultural Issues Offshore

In thinking through the shortcomings of institutional technical communication, we should consider how the diverse nature of ethnic backgrounds and language abilities offers yet another challenge. Oversimplification of workforces leads to another important topic: stereotyping. Too often writers—or designers of safety initiatives—tend to group an entire set of individuals as "people," assuming that all of the targeted audience share the same needs, attitudes, and values. McGee (1975) writes that people are moved "more by maxims and self-interest than by reason and evidence" (pp. 237–238). Ultimately, employees are moved by self-guided incentives rather than anything else. It is also key to remember that one of the main ethnic groups inhabiting offshore platforms and rigs, Indians, comprises a multitude of languages, cultures, and religions within its population; therefore, it is complicated to attempt to guess how "a typical Indian" would think, feel, and react to a given situation when Keralans and other South Indians typically resent speaking Hindi, India's national language. What is likely, and not quite knowable, is how these smaller distinct groups communicate among themselves, taking these safety and professional initiatives and remaking them in their own ways.

A major reason these tactics remain hidden is the extent to which professional power and its structures obscure how these groups communicate. Hall (1976) defines power distance as "the extent to which the less powerful members of institutions and organizations within a country expect and accept that power is distributed unequally" (p. 61). Different countries fall within the spectrum of power distance, from lower to higher. Deference is within the minds of both the employer and the employee, and there are differences in expectation between lower-power-distance environments and higher-power-distance environments. Countries such as those in the Global North lean toward lower power distance, in which it is acceptable for employees to challenge employers about company ideals and employee rights; however, predominantly Asian countries lean toward higher power distance, where there is a far greater disparity or gap between employee and employer. In a working environment such as DO's offshore installations, predominantly low-power-distance employers direct high-power-distance employees; therefore, expectations do not match, and miscommunication ensues. For example, I expect that as much as a supervisor tells his foreman or crew member that "everyone is equal" and to "report your colleagues who violate our safety initiative," this request would be met with skepticism and possibly fear. This is partially due to subordinates' own working experience of high-power distance, in which subordinates are not encouraged to question authority, even if a supervisor exhibits questionable behavior.

Another cultural dimension that is useful to consider here, taken from Hofstede and colleagues (2010), is the notion of high-context versus low-context culture and how it remains a significant factor. Predominantly Asian cultures, which also comprise the four ethnic majorities among the oil platform I visited, operate where nuance carries possibly more meaning than what is actually spoken or written, and those working within high-context cultures must "read between the lines" to comprehend meaning; in contrast, cultures such as those found in the Global North operate under more direct, low-context cultures, where what is spoken or written is typically what is meant. Hall (1976) compares these two working environments:

> It is easier to foresee trouble in low-context cultures because the bonds that tie people together are somewhat fragile, so that people move away or withdraw if things are not going well. In high-context culture, when the explosion comes, it comes without warning. When boundaries are overstepped, they must overstep so far that there is no turning back. It is sheer folly to get seriously involved with high-context culture unless one truly grasps its contexts. This is the danger that the West faces in dealings with the East. (p. 127)

As DO offshore subordinates face a duality of message from "everyone is equal" to possible racial discrimination by their immediate supervisor, they would attempt to "read between the lines" and glean what is truly acceptable and safe company behavior to continue receiving a paycheck, all via their TTC support network.

Individualism versus collectivism is the third cultural dimension taken from Hall's and Hofstede's frameworks that offers space to consider how these marginalized populations communicate on the rig. Hofstede and colleagues (2010) write that "the vast majority of people in our world live in societies in which the interest of the group prevails over the interest of the individual. We call these societies collectivist, which refers to the power of the group" (pp. 90–91). They proceed with a discussion of collectivist societies, in which extended family—whether at work or at home—plays a far more important role than the individual. Predominantly Asian societies rank higher as collectivist, where groups achieve recognition rather than a specific person, and work is often conducted in teams for the greater good of the company "family." On the other end of the spectrum is the individualist ideology in which rewards and recognition for individual achievement are encouraged and promoted. Countries such as those found in the Global North lean more toward individualist ideology than collectivist. Once again, this highlights a difference in expectations between DO supervisors, who come from predominantly individualist cultures, and their subordinates, who come from predominantly collectivist cultures.

Communication is complicated further by an international workforce conforming to a company's specific working culture. Kim and Hubbard (2007) mention that "reducing (mis)understandings or (mis)communication to particular linguistic and/or cultural structures is not only to collude with the existing power relations, including the dominant socioeconomic power, but also to potentially obscure and legitimate the existing power structure" (p. 230). In the Arabian Gulf, hierarchy of power consists of managers from individualist societies, mostly from Denmark or the United Kingdom, and Asian subordinates, who come from typically collectivist societies. Levine and colleagues (2007) mention that "the idea of individualism-collectivism is probably the dominant theoretical perspective guiding cross- and inter-cultural research" (p. 215). Typically, subordinates just want to get along, conform, and assimilate into their work culture. While social orientation impacts the DO communication situation, another factor is communication context. Kim and colleagues (1998) convey the importance of message within context with the Japanese proverb that says, "A nail that stands out gets hammered down" (p. 510). DO's communication issue with offshore subordinates is not only about English language skills but rather a case of poor analysis of message context for this "specific" audience, which consists of primarily South and Southeast Asian nationalities.

Another intercultural issue that contributes to DO's specific situation may most likely be the role of comfort/discomfort. People often feel more comfortable with members of their own race/country/tribe. Vatrapu and Perez-Quiñones (2006) report on a usability study on two independent groups of Indian participants. Results showed that "participants found more usability problems and made more suggestions to an interviewer who was a member of the same (Indian) culture than to the foreign (Anglo-American) interviewer" (p. 156). In breadth of descriptive answers, participants with the Indian interviewer were much more detailed; furthermore, this interviewer from the same ethnic background was included in participants' use of pronouns, where Indian interviewees would say "we" instead of "I," which is further evidence of comfort with the interviewer—an insider rather than an outsider, much like the oil-rig "tribal communities" that DO has mentioned.

Dynamics of a Petroleum (Re-)training Session

Before the meeting began at a five-star hotel's meeting room, Fatboy Slim was played loudly from the speakers as participants entered. There were tables with various juices and appetizers, five large, round tables, and a table with a sign-in sheet. My table featured three Arabs, one South Asian, and me, a North American; I had selected the table on purpose because I was the only Caucasian. Curious looks from other attendees suggested the question, "Why is that man not sitting with his own kind?" I could not determine if my sitting at this multiethnic table raised credibility about me, lowered credibility, or neither. Nevertheless, the Arabs and I bonded over Gulf Arabic pleasantries, and the South Asian and I bonded over my familiarity with his region and culture. Other tables had been entirely occupied by one ethnicity, a peculiar feature of Arabian Gulf life to a foreigner yet a common staple in Arabian Gulf dynamics: one table only had South Asians, another had Southeast Asians, and the final two tables featured Caucasians, which were the two tables closest to the front; these Caucasians were from Denmark, South Africa, the UK, and the USA. An Englishman was the session's discussion leader, and he was flanked by two other discussion leaders, a Scotsman and a Dutchman.

The first part of the session featured a slideshow of oil and gas disasters that could have been prevented. The audience was asked to provide correct answers about how these disasters could have been prevented. When the audience was reticent to respond, the English presenter shouted in an exasperated tone, "Come on people!" Caucasians were the only participants who answered him. Loud air conditioners drowned out a great deal of those who spoke up.

In contrast, the next section of the program was "What management expects from you: (1) If a job is not safe, do not do it or allow workmates to do it. (2) If you see something unsafe, speak up, no matter whom or what. (3) If you are unsure about something, speak up and ask. Do not stop asking until you're satisfied." At one point, the Scottish speaker mentioned that he did not "want to stand here and lecture," but he stood in one spot and lectured for 30 minutes. As he asked the audience safety initiative questions, he addressed the audience's reticence to participate: "Why are you hesitating? Do you feel like the industry says one thing and does another?" I do not know if this was rhetorical as no one volunteered an answer. Frankly, I also did not know if all attendees understood his colloquialisms, his thick accent, and his slang.

An hour later, we listened to a case study of the *Piper Alpha* disaster, which illustrated the importance of refraining from breach of procedures as well as reiterated the importance of implementing risk safety education and culture. "Make safety a priority in work and home life," the English national said. This statement segued into the idea to stay safe at home. The English speaker also said, "Folks, try to be safe; try to remember not to switch on only at work," implying that all those attending should be aware of safety practices both at home and at the workplace. Both leaders mentioned how "changing culture is extremely difficult." Meanwhile, I was still wondering how many of the audience knew what "switch on" meant in this context.

Next, the three speakers showed us a slide of 1930s Rockefeller Plaza with construction workers sitting on a beam that towered above skyscrapers. No safety equipment was evident in this slide. "What do we accept today that we won't accept tomorrow?" asked the English speaker. We provided answers such as "no shoes, safety harnesses, or helmets." The Scottish presenter asked, "Do you think it could happen in this country?" The audience acknowledged that this was possible. At this point in the conversation, the Scottish speaker provided an anecdote that included a questionable description of his maid: "She's Filipino, so she's only this tall," as he raised his hand waist high. After his anecdote, the Scottish speaker asked the audience, "How much illegal hot work is in the North Sea?" According to those in the meeting, North Sea platforms are completely devoid of illegal hot work; in contrast, hot work is quite prevalent in the Arabian Gulf.

My Analysis of (Re-)training Session

This audience of 20 employees of the (re-)training session was numb, uncomfortable, and passive, and they were simply intent on jumping through a mandatory hoop. Audience participation was an absolute minimum.

Attendees from collectivist cultures should feel better by simply working in groups regardless of ethnicity; however, with the dynamic of some attendees requiring additional translations into Arabic, Hindi, Malayalam, and Tagalog, it became clearer why tables were ghettoized. Nevertheless, Hofstede and colleagues (2010) explore fear within we-versus-they experiments, which also reinforce the idea of ghettoization among ethnic lines within a fairly awkward training session for a set of reticent offshore workers: "[P]eople are we-versus-they creatures. It becomes hard for people to change intuitive we-they responses to racial characteristics. Physiological reactions to a we-they situation can be based on any distinction among groups—even that among students from different university departments" (p. 16). At my own table, a Gulf Arab gentleman assisted a visually impaired Gulf Arab colleague and translated everything into Arabic; in the meantime, there were other Arabs at our table who relied on the first Gulf Arab gentleman's Arabic translations, thereby putting TTC in direct practice.

TTC is not only useful in situations where institutional technical communication falters, but it is also clear how useful it was here as the translation strategies into Arabic make clear. Even so, there are clear ways to improve these kinds of workplace situations: diversity in session leaders would have been helpful. After all, the audience is mostly international, and sections such as those in the (re-)training session should encourage ethnic diversity, which in turn would empower its audience members. Sun (2012) discusses the importance of cross-cultural design, the applications of which could be applied anywhere, stating that "in this age of glocalization, cross-cultural design expertise is needed by every researcher, practitioner, or communicator who seeks to have meaningful and engaging conversations for his or her culturally diverse users" (p. 268). Unfortunately, the solitary use of Caucasian leaders possibly reinforced the notion that the "white man knows best" and detracted from the spirit of the (re-)training session. In the least, occasional breakout sessions that feature members of different ethnicities would encourage attendees so that attendees could see relatable examples of appropriate offshore behavior.

The use of specific examples of audio and visual risk communication, namely, approaches to risk and disaster, was very helpful in empowering the audience with its safety message. Ultimately, it is crucial for workers to be reminded and rereminded of the extraordinary amount of risk involved in daily petroleum life. With *Piper Alpha*, the audience could see the cause and effect on the work environment, and with the sobering Charlie Morecraft story, the audience could see cause-effect on the human body. The visual of 1950s Rockefeller Plaza also reinforced how far much of the world has come in terms of safety at work. However, I do not see the point in placing a cross at the top of a projected graphic labeled "Death." Although some may argue that the cross is a universal symbol for death, it is also Christian-centric if not imperialistic, and I believe that the vast majority of the attendees, namely, Sikhs, Muslims, Buddhists, and Hindus, did not associate meaning to it. A final projected photograph showed a blue-collar worker smoking at a construction site. The session leader mentioned, "I don't think that guy in the flip-flops had a good attitude about safety," referring to the worker sitting at the top of a construction site. What if flip-flops were all the worker had to wear? What if the worker simply didn't know they were involved in risky behavior? What if "attitude" had nothing to do with the out-of-context photograph?

Discussion leaders, who are purportedly the pioneers of safety initiatives, must consist of culturally sensitive individuals. Often, lack of intercultural finesse is a result of not understanding other cultures at all, as Hall (1976) alludes to: "People in culture-contact situations frequently fail to really understand each other" (p. 214). The best example of cultural insensitivity during this (re-)training session was one leader's disturbing remark about his Filipino maid's height. A day-long session about a company's safety initiative can be entirely negated by a thoughtless remark like this; hence no safety initiative buy-in, further perpetuation of fear, and more reliance on in-group TTC.

Mixed messages were unintentionally sent during the (re-)training session among leadership. The (re-)training session that I attended reinforced the notion of there being only one appropriate response to crises ("Prisoner, Tourist, or Partner?"), especially those reticent to respond in English as a third or fourth language; and the leaders involved in leading "discussion" could have used more training in eliciting audience responses. While there clearly was important material conveyed in this session, the session appeared to be more of a formality that participants wanted to get over with. The leaders appeared to badger participants for comments when the participants were most likely afraid to get the "wrong answer" and consequently receive a bit of shame or scorn.

Training sessions emphasize leadership by trying to make workers feel comfortable to report unsafe behavior such as shortcuts or illegal hot work. DO shows leadership by deploying their safety initiative for all employees to follow. Ultimately, subordinates show their own leadership as they translate the safety protocol into Hindi, Malayalam, Nepali, and Tagalog so that their colleagues will be able to make informed decisions in the future.

Conclusion

Oil companies employ all kinds of tactics to inform and encourage safe behavior among the onshore and offshore workforce. Genres, if employed correctly, can create a unified front in order to reinforce messages of fraternity, teamwork, collaboration, culture of caring, patriarchal authority, and sustainability. Conversely, the genres also work together to reinforce the company hierarchy/hegemony, fear, miscommunication, cultural differences, and power distance.

It is TTC among translation teams among offshore subordinates that helps in communicating company messages, pivoting around unapproachable supervisors, strategizing work plans, and working through nebulous company messages or cultural marginalization. Whether it is a table of native Arabic speakers translating for each other at an orientation meeting or the translation of training aboard an oil rig among another language community, these tactics make clear that these disempowered populations gain understanding and community through making do on the petroleum rig.

References

Crittenden, G. (2010). *Understanding the initial* Deepwater Horizon *fire*. HazMat Management. http://www.hazmatmag.com/environment/ understanding-the-initial-deepwater-horizon-fire/1000370689

de Certeau, M. (1984). *The practice of everyday life*. University of California Press.

de Certeau, M. (2004). The practice of everyday life: "Making do": Uses and tactics. In G. M. Spiegel (Ed.), *Practicing history: New directions in historical writing after the linguistic turn* (pp. 213–223). Routledge.

Edenfield, A., Holmes, S., & Colton, J. S. (2019). Queering tactical technical communication: DIY HRT. *Technical Communication Quarterly, 28*(3), 177–191. https://doi.org/10.1080/10572252.2019.1607906

Hall, E. T. (1976). *Beyond culture* (1st ed.). Anchor.

Hofstede, G., Hofstede, G. J., & Minkov, M. (2010). *Cultures and organizations: Software of the mind* (Third edition). McGraw Hill.

Kim, D., Pan, Y., & Park, H. S. (1998). *High-versus low-context culture: A comparison of Chinese, Korean, and American cultures. Psychology & Marketing, 15*, 507–521. https://doi.org/10.1002/(SICI)1520-6793(199809)15:6<507:AID-MAR2>3.0.CO;2-A

Kim, M., & Hubbard, A. S. E. (2007). Intercultural communication in the global village: How to understand "the other." *Journal of Intercultural Communication Research, 36*(3), 223–235. https://doi.org/10.1080/17475750701737165

Kimball, M. (2017). Tactical technical communication. *Technical Communication Quarterly, 26*(1), 1–7. http://dx.doi.org/10.1080/10572252.2017.1259428

Leveson, N. (2004). A new accident model for engineering safer systems. *Safety Science, 42*(4), 237–270. https://doi.org/10.1016/s0925-7535(03)00047-x

Levine, T. R., Park, H. S., & Kim, R. K. (2007). Some conceptual and theoretical challenges for cross-cultural communication research in the 21st century. *Journal of Intercultural Communication Research, 36*(3), 205–221.

Lundberg, J., Rollenhagen, C., & Hollnagel, E. (2009). What-you-look-for-is-what-you-find. – The consequences of underlying accident models in eight accident investigation manuals. *Safety Science, 47*(10), 1297–1311. https://doi.org/10.1016/j.ssci.2009.01.004

McGee, M. (1975). In search of "the people": A rhetorical alternative. *The Quarterly Journal of Speech, 61*(3), 235–249. https://doi.org/10.1080/00335637509383289

Nielsen, H. P. (2016). Offshore but on track? Hypermobile and hyperflexible working lives. *Community, Work & Family, 19*(5), 538–553. http://dx.doi.org/10.1080/13668803.2016.1142427

Nielsen, M. B., Glaso, L., Matthiesen, S. B., Eid, J., & Einarsen, S. (2014). Bullying and risk perception as health hazards on oil rigs. *Journal of Managerial Psychology, 28*(4), 367–383. https://doi.org/10.1108/jmp-12-2012-0395

Nielsen, M. B., Tvedt, S. D., & Matthiesen, S. B. (2013). Prevalence and occupational predictors of psychological distress in the offshore petroleum industry: A prospective study. *International Archives of Occupational and Environmental Health, 86*(8), 875–885. https://doi-org.ulm.idm.oclc.org/10.1007/s00420-012-0825-x

Randall, T. S. (2022). Taking the tactical out of technical: A reassessment of tactical technical communication. *Journal of Technical Writing and Communication, 52*(1), 3–18. https://doi.org/10.1177/00472816211006341

Skogdalen, J. E., Utne, I., & Vinnem, J. E. (2011). Developing safety indicators for preventing offshore oil and gas deepwater drilling blowouts. *Safety Science, 49*(8–9), 1187–1199. https://doi.org/10.1016/j.ssci.2011.03.012

Skogdalen, J. E., & Vinnem, J. E. (2011). Quantitative risk analysis offshore—Human and organizational factors. *Reliability Engineering & System Safety, 96*(4), 468–479. https://doi.org/10.1016/j.ress.2010.12.013

Skogdalen, J. E., & Vinnem, J. E. (2012). Quantitative risk analysis of oil and gas drilling, using *Deepwater Horizon* as case study. *Reliability Engineering and System Safety, 100*, 58–66. https://doi.org/10.1016/j.ress.2011.12.002

Sun, H. (2012). *Cross-cultural technology design: Creating culture-sensitive technology for local users.* Oxford.

Vatrapu, R., & Perez-Quiñones, M. A. (2006). Culture and usability evaluation: The effects of culture in structured interviews. *Journal of Usability Studies, 1*(4), 156–170.

Wandekokem, E., Mendel, E., Fabris, F., Valentim, M., Batista, R. J., Varejão, F. M., & Rauber, T. W. (2011). Diagnosing multiple faults in oil rig motor pumps using support vector machine classifier ensembles. *Integrated Computer-Aided Engineering, 18*(1), 61–74. https://doi.org/10.3233/ica-2011-0361

Part 2

Transforming Society

Chapter Ten

Engaging Eurocentric Legacies in de Certeau's Thinking Through a Self-Reflection on "Queering Tactical Technical Communication"

Avery Edenfield and Steve Holmes

As the exigency for this edited collection attests, tactical technical communication (TTC) is an established conceptual lens (Kimball, 2006). In a banal observation, initial scholarship that follows the proposal of a new idea like TTC is obliged to endeavor to create a durable network of practice and citational reference by making arguments for inclusion. Once a concept such as TTC is established, those supporting the network should in turn pause, reflect, and question what elements are working better than others, which is undoubtedly one of the motives the editors had for assembling this collection. In another equally banal observation, no theorist such as de Certeau (1984) is capable of producing an airtight theoretical framework. As Kenneth Burke (1966) observed decades ago, "[A]ny terministic description of reality will only ever be a partial deflection, selection, and reflection of that reality" (p. 45). Thus, every theoretical treatment or conceptual definition will invariably express some degree of a given thinker's respective situated temporal and cultural milieu.

As a case in point, researchers in TC have started to highlight the need to extend the idea of TTC and de Certeau's thinking to clarify the ethical frameworks that motivate a tactical action (Colton et al., 2017). Others have questioned if the initial framing of TTC may have avoided grappling with other aspects of de Certeau's philosophy that could complicate some of the ways that the field has taken up his work (Randall, 2021). For example, de Certeau (1984) offered a productive political and social alternative to fixed Cartesian spatial notions of urban space. Yet, Boyle (2018) has noted spots where de Certeau (1984) offered a privileged, removed, and reflective status to distant urban observers—those at "a supposed distant remove"—on top of tall buildings. De Certeau contrasts the distant observer to those below on the streets who he argues are comparatively more immersed within urban space. Boyle (2018) has questioned this supposed distance because "practices of embodiment and rhetoric also entail dangling one's feet over the side of a building, at a supposed distant remove" (p. 125). However, not all urban practices occur in the streets. Boyle (2018) has explored the recent phenomenon of "rooftopping," which is an emergent vein of urban exploration called "urbexing." In this practice, it is impossible to maintain a separation between distanced observers and urban spatial actors.

While undeniably important to consider, Boyle's (2018) qualification of de Certeau (1984) is hardly debilitating to de Certeau's overall tactical project or to conceptions of TTC. By comparison, postcolonial researchers outside of TC have raised more serious sounding criticism. As we discuss in greater detail below, Rabasa (1993) and Massey (2005) have offered additional critical challenges to certain aspects of de Certeau's (1984) thinking. Both scholars explore different ways in which the latter's writing potentially reinscribes a Eurocentric and (nonwhite, female) other negating conceptual underpinning of space, time, and narrative. In other words, *part of understanding the future of TTC means acknowledging and grappling directly with the less than desirable parts of de Certeau's philosophical approach, especially for those of us who wish to employ TTC for social-justice purposes.*

This chapter explores notable postcolonial criticisms of de Certeau as a starting place to explore the following question: To what extent do technical communicators need to think about de Certeau's problematic elements as we seek to continue utilizing TTC? Can researchers separate the parts that they are able to use productively for social justice, or has the field missed more central ontological and conceptual problems that may work against equity aims? Even if de Certeau (1984) has several thornier issues to qualify, what should we make of the fact that TC scholars, including the current authors of this chapter, have built on concepts such as TTC in the past to do quite progressive things? Despite growing awareness of the politics of citations and structural oppressions, many continue teaching and citing Aristotle and his classical rhetoric concepts such as *ethos, pathos*, and *metis* despite Aristotle's indefensible attitudes toward gender and slavery. Countless postcolonial and decolonial theorists since the 1980s have also pointed to the Eurocentric implications of continually recentering the Greco-Roman rhetorical tradition. With flagship journals such as *Technical Communication Quarterly*'s growing editorial emphasis on attending to the politics of citation, our field is grappling with a similar set of tensions in that our citations and epistemic starting places matter (Walton, Moore, & Jones, 2019) as well as who and how we cite (Agboka, 2014; Itchuaqiyaq & Matheson, 2021). To play the devil's advocate, could the field not come up with ideas similar to TTC by centering an idea like indigenous Woodlands peoples' wampum rhetoric, which Haas (2007) has already demonstrated functioned like a nondigital form of hypertext rhetoric?

To be sure, these are incredibly complicated ethical questions with multiple possible responses and considerations that by no means overlap. Hearteningly, Rabasas (1993) and Massey (2005) both find enduring value in de Certeau's work and call upon his interpreters to explore ways to critique and employ tactics and strategies. This chapter similarly grapples with the Eurocentric nature of de Certeau's thinking and assesses how and in what way it might help researchers critically reflect on the authors' past efforts applying TTC to study queer online DIY health-education practices (Edenfield, Holmes, & Colton, 2019). The authors focused on the tactical behaviors of marginalized individuals (transgender persons), of course, which remains an ethically viable cause.

However, while the authors (with Jared S. Colton) have been recognized for both a Conference on College Composition and Communication and a Nell Ann Pickett Award for the article "Queering Tactical Technical Communication," the authors have realized that they neglected to consider how tactical actions for transgender people are intersectional among others. In reflection, a more ethically replete study of TTC and queer and/or transgender or nonbinary ("trans")[1] users needs to start with an acknowledgment of how enduring a Eurocentric starting place can be even when we are trying to attend to an ethically worthy and clearly marginalized population. What researchers think of de Certeau's work and how they apply it has at least as much to do with the field's ongoing structural racist and colonial tensions as it could also gesture to some insurmountable theoretical contradiction within de Certeau's still useful and relevant writing.

In this chapter, the authors explore and highlight some of the complexities inherent in thinking through Eurocentric bias in hopes that this discussion might aid others who are grappling with similar tensions in their own efforts to connect TTC to social-justice and intersectional concerns. Initially, the authors were interested in exploring an extension of their original study. Specifically, the authors wanted to know whether Black, Indigenous, people of color (BIPOC) trans users would recognize or agree with the use of terms such as *tactics* to understand their own activities of resistance to strategies of control as a way to assess the authors' own uses of the term. However, herein lies the problem. First, the authors will briefly explain some of the tensions within de Certeau's (1984) work through Rabasa (1993) and Massey (2005). Many of de Certeau's problems remain quite minor akin to Boyle's observation about the immersed/nonimmersed binary and are hardly debilitating to the conceptual underpinnings of TTC. As such, the authors situate the broader problem with de Certeau as one in which the field should acknowledge the ongoing use of a European intellectual tradition with respect to critiques by colonial and "decolonial" theory. As Itchuaqiyaq and Matheson (2021) (and, for rhetorical theory's engagements with indigenous rhetoric, Arola, 2017, before them), have argued, decoloniality is an ethically complex process, especially if well-meaning researchers wish to avoid recolonizing indigenous voices, needs, sovereignty, and goals of historically and multiply marginalized colonized peoples. Any time researchers who are outsiders of a community or identity category are "speaking for" the others without enlisting them as collaborators in knowledge production, researchers are at risk of projecting their epistemic norms as theirs. Thus, one problem with our proposed study extension idea lies in the fact that BIPOC trans persons can be very reluctant to participate in academic projects—and rightfully so. History is replete with examples of extraction and exploitation in the name of research.

As of this publication date, the authors can offer a self-reflective narrative—one that should not be confused in any way with a decolonial reading in the sense of Itchuaqiyaq and Matheson (2021), which is engaged in detail below. Instead, this self-reflection is offered as a more partial and suggestive analysis to help the authors as researchers and hopefully others who are interested in serving as allies think through how they can revise and extend certain parts of "Queering Tactical Technical Communication." To do so, the authors feature a self-criticism through a decolonial lens of their original study design. The authors suggest how they would revise the study for future use by comparing their analysis of the online community texts that were used in "Queering Tactical Technical Communication" to similar sets of online tactical texts and YouTube videos by BIPOC trans people. In doing so, they highlight a route forward for themselves to think through how they would start to revise their own past and future uses of de Certeau (1984). The authors' original diagnosis of the strategies facing trans people was at once accurate (and still is), but perhaps only for certain trans user types. Finally, the authors reexamine sites of resistance where trans persons of color locate tactics in relation to powerful institutions with their lives literally on the line, including in online communities and the "cracks between" institutions. The authors raise a problematic in hopes of furthering and extending trans-BIPOC work which is undeniably an ethical good in relationship to enacting social justice in either active or passive senses (Colton & Holmes, 2018).

The Eurocentric Roots of de Certeau's Thinking

The editors of this collection have already provided a literature review of past and recent work in TTC in the introduction. As such, the authors want to begin directly with de Certeau and the issue of Eurocentrism rather than rehearse the litany of excellent past work on TTC in the field.

Layer 1: Eurocentric Legacies Within de Certeau's Writing

There is a wealth of past work on postcolonial spatial theory and the Eurocentric nature of certain spatial conceptions, especially in disciplines such as geography and history. Relevant to de Certeau, Rabasa (1993) has examined de Certeau's use of texts such as Jean de Lery's *Historie of Travels in Brazil* to build his spatial theory (cited in de Certeau, 1988, pp. 225–226; cited in Rabasa, 1993, pp. 46–47). The context occurs coincidentally in a familiar object of study for technical communicators: a caption for an image. Rabasa (1993), who is a colonial historian, has examined the use of cartographic descriptions and visuals in historical references to understand how space is figured. Traditionally, colonizers are figured as active, while colonized peoples are (dis)figured as "blank slates" to be written upon by the gaze of the colonizer. He has written, "Curiously enough a most naive reading recurs in a caption by Michel de Certeau: *"L'explorateur (A. Vespucci) devant l'Indienne qui s'apppelle Amerique"* [The explorer (A. Vespucci) facing the Indian Woman named America], which can "take the reductive categories of savagery and nudity as natural givens" (Rabasa, 1993, p. 41).

Rabasa was puzzled by this discovery. After all, de Certeau follows in Foucault's footsteps and in the footsteps of a number of left-leaning French intellectuals. De Certeau is no uncritical admirer of oppressive frameworks grounded in racism or Eurocentrism. Part of his project lies in directing more orthodox Marxists toward recognizing the roots of potentially revolutionary consciousness in everyday acts. Locating resistance in the everyday practices of life easily and productively extends to the activities of non-European populations who resist oppression but have never read about or heard of "class consciousness" or orthodox Marxist concepts. In various writings about tactics, de Certeau noted that orthodox Marxists were generally more interested in programmatic and institutionalized types of political consciousness.

Nevertheless, de Certeau would have been wise to cast a more critical eye on the older travel narratives that he used to develop his theoretical ideas. Elsewhere, de Certeau was quite critical of a colonial project when warned against "writing that conquers" (1988, p. xxv). Jean van der Straet's historical rendering of Vespucci's first steps on "Latin" America, for example, figured the (soon to be colonized) indigenous other as a synecdoche: "nouva terra" (not yet existing on maps) which is akin, de Certeau has written, to "an unknown body destined to bear the name Amerigo. But what is really going on here is a colonization of the body by the discourse of power" (p. xxv).

Even when it is clear that de Certeau (1984) is no supporter for colonialism, Rabasa (1993) argues that any effort—no matter how small—to figure the nonwhite, non-European other as a blank slate must be relentlessly critiqued. If the authors perform Derridean deconstruction (e.g., taking a marginal element of a text like a caption and demonstrating that this marginalia—the margin—is quite a bit more central to the overall meaning—center—of the text than the writer or interpreters have believed), de Certeau would be far from the first 20th-century European intellectual to invoke tropes of a racialized modern Enlightenment cartography.

Building on Rabasa's (1993) reading, Massey (2005) has taken an additional step and argued that de Certeau's approach to space in these small textual marginalia and asides can reflect a familiar colonial set of binary oppositions between "active historical Europe . . . and an object of the gaze/knowledge," as well as an opposition between time and space (p. 121). Massey (2005) has complained that de Certeau's roots in structuralism lead toward an "insistence on binarism" that can often reduce the complexity of the other even if the former is trying to write about overturning the systems of spatial ordering that produced the latter (Massey, 2005, p. 122). The problematic nature of these sorts of active (Western)/ passive (non-Western) binaries is hopefully by now well understood in TC after the social-justice turn.

Yet, these minor observations are hardly sufficient justification to raise the issue of critiquing de Certeau's conceptual development on the grounds of colonizing the other. It is critical to note that these postcolonial and decolonial criticisms being raised of de Certeau are not central to de Certeau's discussion of strategies and tactic. Consider a different comparison with a more problematic philosopher to reinforce this point. The German philosopher Martin Heidegger (2016) refused to disavow the Nazi party even after it was clear he was aware of its horrific actions. Even worse, his so-called Black Notebooks make it clear that he thought that Jewish peoples were unworthy of an equal ontological category as Germans when it came to theorizing DaSein (Being). Assheuer (2014) has written, "The Jew-hatred in 'Black Notebooks' is no afterthought; it forms the foundation of the philosophical diagnosis" (Assheuer qtd. In Wolin, 2014, para. 5). If it were a mere personal prejudice and not something immanent to Heidegger's entire worldview, then it might be possible for some Heideggerian defenders to have solid ground to stand on. However, passages such as these are damning: "World Judaism is ungraspable everywhere and doesn't need to get involved in military action while continuing to unfurl its influence, whereas we are left to sacrifice the best blood of the best of our people" (qtd. in Olterman, 2013, para. 6). As such, it is understandable if some scholars would seek to avoid citing or even engaging Heidegger's thinking.

The reason for this comparison is because it helps us to understand one way to think through the role that TTC and de Certeau have in the field. In a positive sense, researchers need to raise the need to acknowledge a Eurocentric bias with some of his framings while conceding that de Certeau was trying to push against colonialism, imperialism, and capitalism as an ally. Highlighting this fact, strategies and tactics have been employed by postcolonial theorists to liberatory effects. Spatial orders (strategies) managed colonized peoples, and tactics can be used to help understand sites of resistance and intervention (Gregory, 1994).

Layer 2: Eurocentricity in TC/Rhetoric/the West

Critical-theory discourses employ "immanent critique," following Horkheimer and Adorno (1944), to use the logic of a text or cultural ideology as a weapon to highlight contradictions within that system. Thus, if de Certeau's textual world does not necessarily contain serious or major ethical problems à la Heidegger (2016), the issue immanent critique identifies would point researchers at the broader Eurocentric/Western-centric nature of the field of

TC (and US higher education in general). To put it simply, the existence of a Eurocentric de Certeau problem likely says more about the state of the field in TC relative to decolonial understanding than it does about any major shortcomings in de Certeau's work.

Thus, researchers can ask of de Certeau what some have asked of any European theorist such as Foucault, Derrida, or Cixous: "Why do we keep going back and citing the 20th Century French thinkers like de Certeau, Foucault, etc., in this anti-racist and #metoo era?" Should we instead follow public intellectual and medieval scholar Dan-el Padilla Peralta and argue against teaching the Greco-Roman tradition (Poser, 2021)? Why stop there? Should TC researchers, teachers, and practitioners stop teaching and citing Carolyn Miller or Steven B. Katz and instead reinitiate TC from scratch by featuring related concepts from indigenous thinkers, BIPOC persons, and non-Western methods (as some in our field are)?

Let us imagine that we do not need to reference de Certeau to discuss tactical activity. Indigenous writing, rhetoric, and TC scholars have noted the complicity between the college writing classroom and colonialism: "It is obvious that there is not a university in this country that is not built on what was once native land. We should reflect on this over and over, and understand this fact as one fundamental point about the relationship of Indians to academia" (Gould 1992, pp. 81–82). In a chapter in a landmark collection on indigenous rhetorics (Survivance, Sovereignty, Story), Qwo-Li Driskoll (2015) has coined a term with clear import for tactical thinking: "I have coined the term *decolonial skillshares* to refer to indigenous rhetorics, pedagogies, and radical practices that ask us to continue our rhetorical (visual, material, performative, linguistic, etc.) traditions as indigenous people, to transform cultural memories for both indigenous and nonindigenous people, and to create spaces for all of us to learn and teach embodied rhetorical practices as a tactic of decolonization" (p. 214). For our purposes, note that the word *tactic* occurs within their own description of their aims. It is not a stretch in the least to declare that there is a dominant strategy of control ("intellectual colonialization") that Driskoll's (2015) concept and recommended teaching practices in decolonial skill shares will challenge as a tactical writerly response.

However, if researchers cite or invoke de Certeau to explain skill shares, they may risk overwriting or erasing Driskoll's (2015) efforts. In other words, researchers would unavoidably recenter a colonization of knowledge even if this knowledge could be used to accurately describe some of Driskill's progressive political aims and seek to undermine colonialism and capitalism (as de Certeau's work tries to do). Arguably, most if not all "critical theory" in the 20th century written by formerly colonized and oppressed communities is tactical because such writerly moves have been disavowed by dominant strategies of control that determine who counts as a speaker and what counts as authentic knowledge (of technical communication, of writing). There is a reason that contemporary examples such as Kendi's (2019) discussion of antiracism identifies structural racism as a "zero-sum" game: researchers either actively dismantle structures of racism or they are passively participating in perpetuating racism by being inactive. We could contemplate a very stark binary: every use of a thinker like de Certeau (or Foucault) in TC to describe a non-European context may mean erasing or failing to acknowledge similar work or thinking of other non-Western traditions.

At the same time, other researchers and critical theorists have found ways to work within and against those traditions. Some may still value exploring these ideas as part (but by no means the central element) of a tradition. For example, the Jarilla-Apache philosopher Viola Cordova perceives a great value in Aristotle's virtue ethics. She replaces Aristotle's focus on the "I" of virtue ethics with (by some accounts, essentialist) notion of an indigenous "We" that situates virtues as a relational form. Citing Cordova and others, Itchuaqiyaq and Mathison (2021) have extended Colton and Holmes's (2017) work on (mostly) Western virtue ethics to indigenous virtue ethics grounded in indigenous communal bonds. Purcell (2017) has also noted that Aristotle is interested in individual character formation, while, for comparison, the Aztecs saw virtues grounded in social relations.

In fact, as nonindigenous persons, the authors, like many in a still white-dominated academic institution, must acknowledge their privilege and positionality with this "both/and"' tension. Even seeking to decenter Eurocentric narratives and recognize non-Eurocentric starting places is an ethically fraught process as well. An issue too, lies, in the need to separate a generalized criticism of Eurocentricism and a well-meaning academic effort to "be more inclusive." Rabasa (1993) and Massey (2005), for example, are not connecting decolonial thinking to specific indigenous advocacy forms. In the broader postcolonial theory landscape, the terms *colonial* and *decolonial* can refer to a wide variety of oppressions and resistances. In a powerful article, Itchuaqiyaq and Mathison (2021) have noted that *decolonial* has become a sort of "catch-all" term in TC. They studied and coded most of the recent work in TC that used this term and found that "decolonial [functions] as a euphemism for social justice or humanitarian work" (p. 2). If well-meaning nonindigenous allies merely decolonize a syllabus without connecting their efforts to more substantive forms of advocacy and alliance grounded in supporting land sovereignty issues, then they cannot answer "Yes" to the following question: "Did you design your research/project to be primarily concerned with supporting, respecting, and restoring the sovereignty of Indigenous peoples, lands, and knowledges?" (p. 6)

Following from this example, it would seem like writing a chapter in this edited collection about decolonizing TTC would be at once a worthy and necessary step. Indeed, the authors initially considered writing this type of chapter to update their previous trans research. However, as defined by Itchuaqiyaq and Mathison (2021), this project is not one that the authors, as nonindigenous persons, could hope to undertake in this chapter without working with willing trans indigenous stakeholders, aims, and concerns on land sovereignty issues. For this reason, too, the authors are also not calling from their own positionality to replace *TTC* with other epistemic terms like Driskoll's (2015) above. The authors would not presume to be able to represent these other epistemic traditions adequately on our own. Instead, the authors hope to highlight the complexities of grappling with the Eurocentric implications of their own past work with respect to their own positionalities and the ethical needs of BIPOC trans persons.

New Lines of Research: What Are BIPOC Trans Community's Specific Terms for Circumventing Strategies?

Thanks to Itchuaqiyaq and Matheson (2021), the field knows that a central part of decolonial work as a form of TTC research would be to partner with and interview BIPOC trans users. The authors would need to solicit the latter's active participation and guidance on research aims, agendas, and goals. Unfortunately, the precarity of many research participants who fit into this category means that it can be difficult to locate willing or interested participants from these subject areas. As of the timeline for submission for this chapter, the authors were unable to secure any willing collaborators in the online spaces and communities that they seek to study. As such, this section will enact a research reflection and redesign relative to what publicly available voices the authors are able to use.

In 2017, the authors with Colton ran a survey as an original part of "Queering Tactical Technical Communication." Alongside theoretical discussions, reviewers understandably (and charitably) told the authors to develop the theory and drop the methods/data-collection part rather than attempt to combine the two aspects. Thus, in response to this CFP for the edited collection, the authors thought that they would revisit their original survey and possibly locate any different terms that trans respondents employed. The authors were interested in whether respondents would use terms like *tactical* to describe their implicit tactical activities such as circumventing a lack of healthcare insurance by using Bitcoin to purchase medicine overseas. In turn, the authors hoped to find evidence of different community-/audience-specific ways of describing "tactical" action that we could then use to productively place them in dialogue with de Certeau and TTC work from a trans perspective.

Edenfield has already taken steps to start researching in these directions (Alexander & Edenfield, 2021) not in a decolonial approach, but through the methods of work that acknowledges de Certeau's Eurocentricity. In their intersectional analysis of African American trans healthcare, Alexander and Edenfield (2021) have argued that African American historic use of folk medicine and trans DIY tactics should be understood as acts of self-care and resistance to oppressive and even abusive medical treatment rather than as acts of patient noncompliance or, worse, ignorance. They found productive congruence between these two distinct and connected communities: a reliance on 91) elders and other trusted people for vetting of information, and 92) online forums and media channels to find and vet information, both activities in response to deep historical and current trauma and abuses. Importantly for this chapter's reflection, they further identify dismissal of these self-care strategies as Euro- and cis-centrism.

As the authors have been reflecting on our own efforts in writing this chapter, they have realized a Eurocentric bias to our survey design: none of their IRB questions differentiated the experiences of BIPOC trans users. This self-realization point is critical. While the authors were pursuing an ethically just cause—exploring the tactical activities of trans users—they reinscribed an ethically unjust marginalized universal subject: a nonraced trans user. Their survey design presupposed a sort of "universal" subject at odds with the actual "singular" ontological framework of "Queering Tactical Technical Communication." The authors' survey results, which they hope to publish on at some point in the future in a TC journal, remain valuable, but this oversight highlights the ongoing need to grapple with Eurocentric roots. How might that study look different, for example, if the authors used Kendi's (2019) antiracist framework as a heuristic or another intersectional approach grounded in racial difference?

To answer these questions with respect to the set of concerns that the authors have raised, consider answers given by BIPOC trans persons on another tactical effort: YouTube transition documentary videos/recordings.

For any medical operation in an institution, doctors or healthcare providers provide medical documentation both for their own institutional records and for legal and insurance purposes. However, some online trans users have (loosely) appropriated this institutional record keeping with hybrid transition documentary videos. While the purposes differ widely, one element they share in particular is a commitment best summarized in a post by Reddit user core0746 (2021): "[A]s a Mexican trans man I've only ever gotten information about transgenderism from the internet and YouTube. But I've noticed the only big and known trans youtubers are white. I only know some poc trans women like Kat blaque and Eden the doll but I haven't found any pic transmen on youtube." This user question highlights some of the unwitting oversights of the author's previous work. If any anonymous online forum as a potential survey site does not reveal much—if any—explicit acknowledgment of the racial discourses of users, then it's a fair assumption that a trans survey population—especially if we fail to ask for voluntary ethnical background information—is probably white.

Acknowledging the difference between trans users' tactical aims and needs reveals an entirely different set of "hacks" to the process of transitioning. For example, transition records produced by YouTube for male users who were assigned female at birth frequently expressed concerns over growing facial hair with an explicit eye toward noting how their experiences differed from white trans persons. A YouTuber self-identified with a Chinese background named "Finnegan" (2017) declared in one YouTube video record, "I'm getting a lot more hair of course. This is for Chinese person and more hair compared to a Caucasian person [which] is very different." A popular African American trans YouTuber, "Matt Turner," has an entire series of informal medical reports posted about "The Black Man's Beard Chronicles." As a tactical medical genre, Turner is documenting his/their yearlong medical journey on YouTube (Turner, 2021a). Turner has described his/their work as a form of documentary activity: "I'm making this series just to document my beard progress and journey using minoxidil" (Turner, 2021b). In this video series, Turner has given advice on how he/they picked his/their doctor.

While they are not as common as Caucasian-dominated trans social-media forums or issues, there are a small number of BIPOC-oriented social-media threads. As a case in point, Turner also moderates the r/transpoc and r/asktransgender Reddit threads. On this forum, another Bangladeshi user "Mayagirli" (2021) posted about facial hair removal being culturally specific and, in turn, sought tactical advice for how to proceed:

> I've been wanting to get rid of my facial hair for a long time, I was using an at home IPL Laser that had a lot of great reviews but it's done nothing for me and even burnt me a few times. I've been using it for I'd say half a year and was just waiting for some kind of results, nothing has changed, so I am now strongly considering professional laser or electrolysis treatments.
>
> [H]owever, a lot of forums and advice out there on whether to use electrolysis or laser for transfems is whitecentric, a lot of white people's answers and experiences. I'm Bangladeshi, my skin tone is equivalent to Monet X Change lol, so I was wondering if any POC's have any experience or advise on using either laser or electrolysis, any help would be so so so so appreciated! <3

These posts, however, reiterate the same refrain: BIPOC representation and, therefore, ethnically specifically technical knowledge about certain aspects of transitioning is lacking among YouTubers and other social-media driven sources of information.

Returning to Matt Turner's videos, in Turner's discussion of the strengths and weaknesses of a prosthetic penis (known as a "packer"), along with reviewing the material, aesthetics, functions, and features (a review clearly situated in technical definition conventions), Turner discusses the difficulty of color-matching darker skin tones (Turner, 2020). Commenters agree, one stating, "Being a ftm of color I'm always let down by the color options" (Saadiq Ahmed Edmond, 2021). Saadiq Ahmed Edmond (a real name or username) creates transition videos to engage this nexus of concerns. One comment on one video, "GMP Review/T-gel update/3 months post-op," echoed Saadiq's comment on Turner's video: "I have such a hard time finding black ftm guys I'm so grateful for ur vids fr" (Miller, 2020). In one video, Turner showcases other trans YouTubers, including several Black transmen, in a way to bring more attention to lesser-known trans creators.

Lack of representation and visibility is not only the case for trans men of color. Among videos on breast forms, prosthetic breasts sometimes used by trans women, there are few YouTube reviews for trans women of color. One YouTube channel with 147,000 subscribers hosts a video which leads to an Amazon product page for "Vollence Black Silicone Breast Forms Dark Fake Boobs for Transgender Mastectomy." One reviewer has stated, "I'm keeping them tho since it's INCREDIBLY difficulty [*sic*] to find these at a bargain . . . for black folk specifically" (Phoenix, 2020). These examples of the availability and representation of color-matched packers, breast forms, stand-to-pee (STP) devices, and other gender-affirming prosthetics unveil the persistence of the racially unmarked trans subject as a white trans subject. Furthermore, these videos and comments cast light on trans-BIPOC uses of technical communication, including technical definitions and documentation. In the previously mentioned Turner (2020) video review of a packer, commenters ask about what color he/they ordered. Turner replied, "I believe it was M16. I had the custom color done and I wanted the M16 with the undertone of M15 but I think it arrived as just the pure M16 color" (2021). "Henry in training" responds, "I had the exact opposite problem, I ordered m15/16 and it was too light haha" (2021). In doing so, Turner brings attention to one example of BIPOC trans uses of TTC as a still-productive lens for helping us locate the informal tactical actions of marginalized users.

Conclusion: Revisions Toward Future Trans-BIPOC Scholarship in TC

The above narrative and reflection consider our past research and uses of de Certeau and TTC in light of reconsiderations of decoloniality, influenced by current scholarship. Here are some ways the authors could expand/get around the limitations of their original surveys.

First, at the time they did not consider intersectional gender studies (Crenshaw, 1991). If the authors were going to go back and explore TTC again using this study, they could revise survey questions to consider a range of intersectional identities in terms of ethnicity, cultural background, class, ability, and other identities. For example, their original questions assumed out-of-the home and in-the-office employment. The authors did not consider people who worked from their car or worked from their home. In this way, the authors made assumptions about workplaces and employment that were unintentionally exclusionary. In an updated study, the authors would add flexibility to questions regarding employment, living situations, and health-care access. Drawing from Alexander and Edenfield (2021), when asking about expertise and trusting information, the authors could include questions about elders and trusted community members, accounting for past and current experiences of trauma and harm at the hands of medical providers. Further, drawing from Jones (2017) and Jones and Walton (2018), the authors would need to include personal narratives collected through interviews. Interviews would also allow us to consider the singularity of individual experiences, an important facet of trans research that we called for in our past work. The authors also recognize that our survey was available in only one language (English), a fact that limited responses. A revised study would include translations. Furthermore, study recruitment would work in partnership with trusted community groups with a goal of collaboration, that is, a good-faith effort to create research products that support a community partner's mission as called for by other researchers, including Cushman (1998), Eva and Patriarcha (2012), Rose and colleagues (2017), Strohmayer, Clamen, and Laing (2019); Walton, Zraly, and Mugengana (2015), among many others. The authors also need to consider the risks (online and off) to participants as called for by Cheek, Clem, and Edenfield (2021) and de Hertogh (2018). The authors' analysis should ideally draw from BIPOC trans research including Gossett (2017), Gill-Peterson (2014), and others. And finally, the distribution of findings and deliverables should be shared with stakeholders in forms that are actionable and that support community missions.

These changes, of course, are just starting places for exploring post- and decoloniality in relation to de Certeau's work. As such, the authors invite scholars from a range of positionalities and identities to join them in similar self-critique of their past research with an eye toward producing more ethical and just futures for trans and queer TPC collaborators.

Note

1. Here and throughout, we use *trans* as a broad term intended to include the experiences of transgender, agender, genderqueer, Two-spirit, hijra, nonbinary, and other gender-expansive identities. We also include persons with and without dysphoria and those who seek or are medically transitioning and those who do not seek it.

References

Agboka, G. Y. (2014). Decolonial methodologies: Social justice perspectives in intercultural technical communication research. *Journal of Technical Writing and Communication, 44*(3), 297–327. doi: 10.2190/tw.44.3.e

Alexander, J-J., & Edenfield, A. C. (2021). Health and wellness as resistance: Tactical folk medicine. *Technical Communication Quarterly, 30*(3), 241–256. https://doi.org/10.1080/10572252.2021.1930181

Arola, K. L. (2017). Composing as culturing: An American Indian approach to digital ethics. In K. A. Mills, A. Stornaiuolo, A. Smith, & J. Z. Pandya (Eds.), *Handbook of writing, literacies, and education in digital cultures* (pp. 275–284). Routledge.

Boyle, C. (2018). *Rhetoric as a posthuman practice.* Pittsburgh: Pittsburgh University Press.

Burke, K. (1966). *Language as symbolic action: Essays on life, literature, and method.* University of California Press.

Cheek, R., Clem, S. & Edenfield, A. C. (2021). Digital research and qubit ethics: Trans* vulnerability and transparency activism. *Proceedings of the ACM Special Interest Group on the Design of Communication Conference (SIGDOC 2021),* 101–107. https://doi.org/10.1145/3472714.3473628

Colton, J. S., & Holmes, S. (2018). A social justice theory of active equality for technical communication. *Journal of Technical Writing and Communication, 48*(1), 4–30. https://doi.org/10.1177/0047281616647 80

Colton, J. S., Holmes, S., & Walwema, J. (2017). From NoobGuides to #OpKKK: Ethics of anonymous' tactical technical communication. *Technical Communication Quarterly, 26*(1), 59–75. doi: 10.1080/10572252.2016.1257743

Core0746 (2021). *POC transgender youtubers?* [Online forum post]. Reddit. https://www.reddit.com/r/transpoc/comments/hochv4/poc_transgender_youtubers/

Crenshaw, K. (1991). Mapping the margins: Intersectionality, identity politics, and violence against women of color. *Stanford Law Review, 43*(6), 1241–1299. doi: 10.2307/1229039

Cushman, E. (1998). *The struggle and the tools: Oral and literate strategies in an inner city community.* State University of New York Press.

de Certeau, M. (1984). *The practice of everyday life.* University of California Press.

de Certeau, M. (1988). *The writing of history.* Columbia University Press.

de Hertogh, L. B. (2018). Feminist digital research methodology for rhetoricians of health and medicine. *Journal of Business and Technical Communication, 32*(4), 480–503.

Driskill, Q. L. (2015). *Decolonial skillshares.* Utah State University Press.

Edenfield, A. C. (2020). Managing gender care in precarity: Trans communities respond to COVID-19. *Journal of Business and Technical Communication, 35*(1), Online Publication. https://doi.org/10.1177/1050651920958504

Edenfield A. C., Holmes, S., & Colton, J.S. (2019). Queering tactical technical communication: DIY HRT. *Technical Communication Quarterly, 28*(3), pp. 177–119. doi: 10.1080/10572252.2019.1607906

Evia, C., & Patriarca, A. (2012). Beyond compliance: Participatory translation of safety communication for Latino construction workers. *Journal of Business and Technical Communication, 26*(3), 340–367. doi: 10.1177/ 1050651912439697

Finnegan (2017, Nov. 17). *FTM: Chinese facial hair* [Video]. YouTube. https://www.youtube.com/watch?v=nf9uPXYMya4

Gill-Peterson, J. (2014). The technical capacities of the body: Assembling race, technology, and transgender. *Transgender Studies Quarterly, 1*(3), 402–418.

Gossett, C. (2017. Blackness and the Trouble of Trans visibility. In R. Gossett, E. A. Stanley, & J. Burton (Eds.), *Trap Door: Trans Cultural Production and the Politics of Visibility* (pp. 183–191). MIT Press.

Gould, J. (1992). The problem of being 'Indian': One mixed-blood's dilemma." In S. Smith & J. Watonson (Eds.), *De/colonizing the subject: The politics of gender in women's autobiography* (pp. 81–87). University of Minnesota Press.

Gregory, D. (1994). *Geographical imaginations.* Wiley-Blackwell.

Haas, A. M. (2007). Wampum as hypertext: An American Indian intellectual tradition of multimedia theory and practice. *Studies in American Indian Literatures, 19*(4), 77–100. https://www.jstor.org/stable/20737390

Haas, A. M. (2012). Race, rhetoric, and technology: A case study of decolonial technical communication theory, methodology, and pedagogy. *Journal of Business and Technical Communication, 26*(3), 277–310. https://doi.org/10.1177/1050651912439539

Heidegger, M. (2016*). Ponderings II–VI: Black notebooks 1931–1938.* Indiana University Press.

Henry in Training (2021, Feb.). *ReelMagik Honest Product Review | was it WORTH IT | ftm/ftn* [Video]. YouTube. https://youtu.be/KmJQgE5LWz4

Horkheimer, T., & Adorno, M. (1944). *The culture industry: Mass enlightenment as deception.* Marxists.org. https://www.marxists.org/reference/archive/adorno/1944/culture-industry.htm

Itchuaqiyaq, C. U., & Matheson, B. (2021). Decolonizing decoloniality: Considering the (mis) use of decolonial frameworks in TPC scholarship. *Communication Design Quarterly Review, 9*(1), 20–31. https://doi.org/10.1145/3437000.3437002

Jones, N. N. (2017). Rhetorical narratives of black entrepreneurs: The business of race, agency, and cultural empowerment. *Journal of Business and Technical Communication, 31*(3), 319–349. https://doi.org/10.1177/1050651917695540

Jones, N. N., & Walton, R. (2018). Using narratives to foster critical thinking about diversity and social justice. In A. M. Hass, & M. F. Eble (Eds.), *Key theoretical frameworks: Teaching technical communication in the twenty-first century* (pp. 241–267). Utah State University Press.

Kendi, I. X. (2019). *How to be an antiracist.* One world.

Kimball, M. A. (2006). Cars, culture, and tactical technical communication. *Technical Communication Quarterly, 15*(1), 67–86. doi: 10.1080/10572252.2017.1259428

Massey, D. B. (2005). *For space.* Sage.

Mayagirli (2021, May 29). *Electrolysis or laser . . . reddit* [Online forum post]. Reddit. https://www.reddit.com/r/asktransgender/comments/nm9vdg/electrolysis_or_laser_for_facial_hair_removal_for/

Miller, J. (2021, n.m, n.d.). *Re: GMP Review/T-gel update/3 months post-op* [Video]. YouTube. https://youtu.be/886T2LeU5s8

Olterman, P. (2014, Mar. 12). Heidegger's "Black Notebooks" reveal antisemitism at core of his philosophy. *The Guardian Online.* https://www.theguardian.com/books/2014/mar/13/martin-heidegger-black-notebooks-reveal-nazi-ideology-antisemitism

Phoenix. (2020, July 31). *They'll suffice.* Amazon. https://www.amazon.com/gp/customer-reviews/R342VGO5Q049PA/ref=cm_cr_dp_d_rvw_ttl?ie=UTF8&ASIN=B0876N9XBN

Poser, R. (2021, Feb. 2). He wants to save classics from whiteness. Can the field survive? *The New York Times* online. https://www.nytimes.com/2021/02/02/magazine/classics-greece-rome-whiteness.html

Purcell, L.S. (2017). Eudaimonia and neltiliztli: Aristotle and the Aztecs on the good life. *APA Newsletter on Hispanic/Latino Issues in Philosophy, 16*(2),10–21.

Rabasa, J. (1993). *Inventing America: Spanish historiography and the formation of Eurocentrism* (Vol. 11). University of Oklahoma Press.

Randall, T. S. (2021). Taking the Tactical Out of Technical: A Reassessment of Tactical Technical Communication. *Journal of Technical Writing and Communication, 52*(1), Online Publication. https://doi.org/10.1177/00472816211006341

Rose, E. J., Racadio, R., Wong, K., Nguyen, S., Kim, J., & Zahler, A. (2017). Community-based user experience: Evaluating the usability of health insurance information with immigrant patients. *IEEE Transactions on Professional Communication, 60*(2), 214–231. 10.1109/TPC.2017.2656698

Saadiq Ahmed Edmond. (2021, Feb.). *Re: ReelMagik Honest Product Review | was it WORTH IT | ftm/ftn* [Video]. https://youtu.be/KmJQgE5LWz4

Strohmayer, A., Clamen, J., & Laing, M. (2019). Technologies for social justice: Lessons from sex workers on the front lines. *Proceedings of the 2019 CHI Conference on Human Factors in Computing Systems,* 1–14

Turner, M. (2020, Dec. 18). *ReelMagik Honest Product Review | was it WORTH IT | ftm/ftn* [Video]. YouTube. https://youtu.be/KmJQgE5LWz4

Turner, M. (2021a). *Vlogs. Fitness.* YouTube. https://www.youtube.com/channel/UC_LpKF4BUDwUBIsCq9O4lzQ

Turner, M. (2021b, Feb. 5). *The black man's beard chronicles.* YouTube. https://www.youtube.com/watch?v=ZYEsItkrc1s&list=PLXABEjM69svzNcZ1Fi0OZjGPeMEWc7z8E&index=8

Walton, R., Zraly, M., & Mugengana, J. P. (2015). Values and validity: Navigating messiness in a community-based research project in Rwanda. *Technical Communication Quarterly, 24*(1), 45–69. https://doi.org/10.1080/10572252.2015.975962

Wolin, R. (2014). Heidegger's black notebooks. *Jewish Review of Books.* https://jewishreviewofbooks.com/articles/993/national-socialism-world-jewry-and-the-history-of-being-heideggers-black-notebooks/

Chapter Eleven

DIY Instructions in *BIKINI KILL*

A Feminist Historiographical Approach to Tactical Technical Communication

John L. Seabloom-Dunne

Feminist theory is a rich and varied intellectual tradition concerned with gender, power, experience, and expression. Its historic and ongoing contributions to the field of technical communication are well known. Among many others, they include challenges to technical communication's commitment to scientific positivism (Lay, 2004), expansions of the kinds of knowledge considered appropriate for the field (Durack, 2004), and a renewed focus on the ways in which technical communication functions within various social and political contexts (Walton et al., 2019).[1] Feminist theory has played a crucial role in developing the topics, methodologies, and purposes that structure contemporary technical communication scholarship.

In this chapter, I argue that we can continue to draw from feminist theory as we study examples of tactical technical communication (TTC): technical documentation that cuts against the grain of professional institutions. I contend that a feminist historiographical methodology is particularly well suited to investigating historical examples of TTC. This chapter first outlines feminist historiography's key features and contributions to rhetorical theory writ large. It then demonstrates the methodology's purchase through an analysis of a text that both embodies and contests technical communication conventions, the feminist punk zine *BIKINI KILL*.

Feminist Historiography and Technical Communication

Feminist historiography, as described by Jessica Enoch (2013), often entails the leveraging of "gender theory to interrogate the relations of power operating inside the [historical] tradition that defines and naturalizes legitimate (masculine) and illegitimate (feminine) rhetorical theory, practice, and knowledge" (p. 67). As historiography is wont to do, this interrogation reveals history's constructed nature and the hollow claims of neutrality and objectivity found in its study. The history of rhetoric gives way to histories of rhetorical practices, theories, and figures. In this chapter, I organize my analysis around this distinction between legitimate and illegitimate rhetorical practices. Attending to this aspect of a rhetorical practice can help scholars understand how its historical position (or absence thereof) has come to be. This approach is especially useful when considered alongside similar axes of legitimacy that include race, class, disability, gender identity, and sexual orientation. All of these factors shape which rhetorical practices have a place and which are cast adrift. Not being afforded a domain of their own, examples of TTC are often by definition illegitimate and insurgent, working at cross purposes to otherwise dominant institutions. *BIKINI KILL* is one such example.

In addition to issues of legitimacy, feminist historiography has also expanded the range of people who are considered to be rhetoricians (Campbell, 1993; Glenn, 1997; Logan, 1999). It has been particularly successful in this respect when recovering historical women rhetors who have been written out of the canon. Scholars of feminist historiography have also critiqued the very concept of a rhetorical canon, seeking other ways of understanding rhetoric's past and future (Biesecker, 1992; Enoch, 2013; VanHaitsma, 2016). Feminist historiography enriches a field of study by demonstrating how its historical conventions and traditions are constructed through the process of scholarship. In the case of TTC, it understands the field's already-constructed history and accepts unconventional texts as meaningful contributions to technical communication.

Zines, *BIKINI KILL*, and the Riot Grrrl Movement

BIKINI KILL, the central example of this chapter, is a zine published by Kathleen Hanna and the other members of the punk band from which it takes its name: Billy Karren, Kathi Wilcox, and Tobi Vail. Published in

1991, the first volume of *BIKINI KILL* presents itself as a text that aims to teach strategies for community safety—both at punk shows and in the world at large—and explore revolutionary feminist politics. The second issue continues in a similar vein, including the now famous riot grrrl manifesto that would eventually spread to international audiences. Both issues are made up of a blend of lyrics, essays, art, annotations, and instructional materials. Although originally distributed at musical performances, these zines were recopied, remixed, and redistributed through informal networks in the months and years that followed.

Many of *BIKINI KILL*'s distinct characteristics arise from its identity as a zine in the early 1990's. As a term, *zine* resists neat categorization. In his history of the medium, Stephen Duncombe (2017) defines zines both as "noncommercial, nonprofessional, small-circulation magazines which their creators produce, publish, and distribute by themselves" and as "little publications filled with rantings of high weirdness and exploding with chaotic design" (pp. 4–9). Although individual and amateur publishing communities have existed for centuries in various forms, the context of zine publication surrounding *BIKINI KILL* rests in the second half of the 20th-century counterculture movements and communities in the United States. Zines would eventually play a central role in these communities: acting as creative outlets for personal expression and as an ad hoc infrastructure for many people who found themselves at odds with mainstream society.

The vast majority of these zines were not the polished products of well-coordinated groups of writers, artists, designers, and publishers. They were produced mostly by individuals using cheap, scavenged, or stolen materials. One zine editor reports: "[M]y roommate . . . works the night shift at Kinko's. She's been copying for people for free for years. In some ways I would say that she keeps the Bay Area zine scene going" (as cited in Duncombe, 2017, p. 88). Zinesters formed networks of PO boxes and mailing addresses where they would share their work with one another, sending each other new issues of their own zines or letters responding to some story, exclamation, or realization. Zines would often conclude with entreaties for responses or with an imperative that the reader make their own zine and take an active role in the ecosystem. While there were certainly people who would only consume zines as reading material, the community as a whole encouraged direct participation in the creation and circulation of these texts.

The characteristics that made zines so popular during their initial explosion onto the countercultural scene in the United States remain compelling for us today. This network of publishers, interlocutors, critics, and correspondents produced a rich and varied body of texts, many of which embody the traits associated with TTC: positioned outside of the conventional processes of publication, composed with makeshift and repurposed tools, leveraging personal experience as a source of expertise, and critiquing the dominant cultural forms of the day.

These zines were produced outside of any single institution, growing out of communities that were self-regulating in content and culture. In them, a zine's continued existence was largely determined by the tenacity of its authors. Authenticity became a core value of zines at the time, and there was a sense that personal expression had a value and importance all its own. In some ways, the content and quality of a zine was secondary to its function as an honest expression of its author's life and perspective. That being said, this emphasis on authentic individual expression did not preclude zines' engagement with broad cultural and political systems as topics of discussion and critique. Indeed, it is one's personal experiences with these systems that often necessitates such engagement. In *BIKINI KILL*'s case, its authors and readers' experiences of misogyny and the systems of oppression that structure the broader culture were one and the same. *BIKINI KILL* grew out of a particular context experienced by the members of the band in the punk scene in the Pacific Northwest, and it aimed to address patriarchy as a societal system of violence and repression.

BIKINI KILL's Technical Function and Tactical Identity

In the sections to follow, I explore and affirm *BIKINI KILL*'s fitness as an example of TTC. I first explore both the zine's technical function and its tactical identity. I then shift to a discussion of those characteristics that hold particular interest for scholars and practitioners of technical communication. These characteristics include the do-it-yourself (DIY) practices of composition and circulation that shape the zine, the zine's partial and provocative forms of instruction, its embrace of personal experience as a meaningful site of knowledge production, and its consistent rejection of cultural and professional norms.

In establishing *BIKINI KILL*'s technical function, K. T. Durack's definition of technology proves helpful for laying out the key concepts at play. Durack (2004) effectively accounts for the capacious nature of technology as an organizing concept within the field of technical communication, writing, "Technology refers equally to knowledge, actions, and tools: it is (for example) a network of constructed waterways, the knowledge of when and how to irrigate fields, and the entire set of human actions that comprise this method for farming" (p. 41). What unites these otherwise dissimilar components is the employment of systems—be they conceptual or material—to some worldly end.

Technical communication then, understood as the rhetorical acts that address, shape, and are shaped by these technologies, must be similarly broad in its topics and similarly concerned with the knowledge, actions, and tools that make up the systems of our world. *BIKINI KILL* aims to analyze patriarchy as a social and political system, transform this system through organized collaborative ongoing action, and support this action in no small part through instructional materials. One member of the band writes, "We Made this fanzine together. and we all play Music together. And sometimes this is all very hard cuz this world doesn't teach us how to be truly cool to each other and so we have to teach each other" (*BIKINI KILL GirlPower*, 1991, p. 1). With this goal in mind, we can imagine how these zines are taken up by readers not as documents to be interpreted but as tools for learning that facilitate further activities—a demonstrably technical function. At the very least, the cover of *BIKINI KILL*'s first issue defines itself as "a color and activity book." It uses the language of action, composition, and creation instead of interpretation or reflection.

BIKINI KILL also exhibits the tactical characteristics described by Michel de Certeau (2011) in that it cannot count on its own spaces, tools, and conditions. It inhabits a place that "belongs to the other" (p. xix). *BIKINI KILL* works to reshape the world with tools not of its own making, expressing the classical technical communication concern for instruction in the context of a tactical struggle shaped by conditions predetermined and hostile to its flourishing. It orients itself against dominant cultural models of knowledge and composition, and this orientation is born out through its ad hoc composition, its distinct forms of instruction, its embrace of personal experience as valuable knowledge, and its explicit rejection of cultural and professional norms. This blend of characteristics positions *BIKINI KILL* as a tactical document, and *BIKINI KILL*'s focus on the systems of cultural and material production demonstrates its technical character.

DIY Practices of Composition and Circulation

Perhaps the most obvious expressions of these zines' tactical identity lie in the materials with which they have been constructed and the methods by which they have been distributed. As a medium, zines have a different set of constraints than those that structure conventional technical communication. For instance, in its initial publication, it is easy to imagine *BIKINI KILL* having a limited circulation. This zine has a narrow focus in some respects, revolving around the music and perspective of a single band in a fairly niche subculture. It is produced by only a handful of people, none of whom are professional layout artists, editors, or publishers. *BIKINI KILL* is in many ways an exemplar of the amateur fanzine genre of the early nineties. Although its content and message would eventually find an expansive and international audience, *BIKINI KILL* is composed for a particular purpose in a by-then well-established genre. As a result, this zine does not attempt to account for a broad audience. It does not need to be easily edited, updated, or reformatted. Nor does it need to abide by an institutional style guide or account for those legal liabilities that come with any stable institutional identity. Instead, this zine is composed and circulated under the constraints of limited materials (whatever the Bikini Kill band members could scrounge, purchase, or pilfer) and limited production capacity (individual printers and copy machines). Found materials collaged together make up a significant proportion of these first two issues of *BIKINI KILL*. Original essays and frequent annotations intersperse these repurposed materials, directly presenting the perspectives of the authors.

However, even this distinction blurs in the final stages of composition and distribution. Like the vast majority of zines at this time, the first two issues of *BIKINI KILL* were predominantly distributed in the forms of black-and-white photocopies of an original document. While *BIKINI KILL* makes full use of word processors, handwritten annotation, and illustration, as well as its wide variety of collaged material, these various media are flattened in the act of copying and compiling the zine for final distribution (figure 11.1). The extent to which content is originally composed is not always clear, and there is little need for such clarification in any case. Collaged excerpts from zines and other media, art pieces, and typed essays blend before the reader's eyes and take on new meanings. Regardless of whether materials were original contributions, repurposed content from other zines, or excerpts drawn from entirely distinct sources, they are presented without distinction.[2] These materials have been transformed through the act of composition into something new, original, and in the eyes of the community, authentic.

Just as *BIKINI KILL*'s blend of content neither encourages nor supports an ongoing practice of citation and reference, the processes of circulation associated with zine's medium ensures that its readership similarly contrasts

Figure 11.1. Excerpted page from *BIKINI KILL*, volume 1. *Source*: *Bikini Kill Zine 1*. Photo by the author from their private collection.

conventional alternatives. These zines may have been available at the band's shows, reviewed or referenced in other similar publications, or mailed out upon request. However, these methods of circulation would have none of the reach or consistency associated with mailing lists and newsstands. Magazines, newspapers, and other periodicals with meaningful financial backing and regular publishing schedules dominate these conventional guarantors of readership. *BIKINI KILL* would have instead depended on its readers and interlocutors to circulate the zine on their own initiative, recommending it through word-of-mouth or sharing and independently producing copies of the zine. While this system remains inscrutable to any circulation metric that depends on tracking sales numbers, impressions, or citations, it does help to ensure that those readers who do find *BIKINI KILL* do so through their own initiative and within the community where the zine was first composed.

Partial and Provocative Forms of Instruction

As a topic, TTC is concerned with teaching, especially the teaching of subjects that the existing dominant institutions either neglect or discourage. Technical-communication scholars approaching this topic have engaged a wide range of instructional materials that include independent car maintenance and construction (Kimball, 2009), covert bomb making in domestic spaces (Sarat-St. Peter, 2017), and DIY methods of hormone therapy (Edenfield et al., 2019). These materials differ significantly in topic, purpose, and scope, but they share a rejection of the status quo and a belief that new forms of instructional materials are needed in order to achieve their ends.

This approach to instruction can also be found in *BIKINI KILL* (*BIKINI KILL: A COLOR AND ACTIVITY BOOK*, 1991). In the final pages of the zine's first issue, on a single relatively sparsely adorned page, the zine presents its readers with a thought exercise. It asks its reader, "Can You Run [. . .] For Your Life?" and suggests that they "practice deciding on a self-defense strategy by reading through the following situations and planning what you would do" (p. 24). The zine then lists a series of nine threatening situations in which readers may find themselves, ranging from being approached by men at a bar while preferring to sit alone to hearing a previously unknown intruder in the home while bathing. No additional advice, recommended steps, or guidance is provided. The reader is presented with these situations as prompts for reflection and preparation, not as the first step in a comprehensive model for response that the reader could follow.

While the zine commands that the reader "practice," the nature, scope, and results of this activity are left ambiguous. Incomplete instructions of this sort are known to appear in examples of TTC, especially when the material discussed is potentially dangerous, illegal, or particularly dependent on inducing audience participation (Sarat-St. Peter & St. Peter, 2020). In *BIKINI KILL*, there is no clear benchmark for how much practice would be sufficient to guarantee the reader's safety. Nor does this series of situations conclude with a mute acceptance that there is nothing to be done. Instead, it demonstrates the necessity of readers developing their own capacity for action. The surrounding content in *BIKINI KILL* lends this section a crucial context in that it maintains a sharp focus on the potential for social transformation through the production of new forms of knowledge and the connections between individual actions and community building.

Partial and provocative instructional materials appear throughout both issues of this zine. A striking example of this practice can be seen in *BIKINI KILL GirlPower*, on page 27 (1991). There, still intelligible despite repeated copying, a diagram of a DIY graffiti rig is positioned alongside several annotations (figure 11.2).[3]

This diagram exhibits many of the characteristics shared by its more conventional counterparts: a simplified graphic style that emphasizes each component's function, numbered labels, and a depiction of the mechanism's

Image 11.2. DIY graffiti rig. *Source: BIKINI KILL*, issue 2. Photo by the author from their private collection.

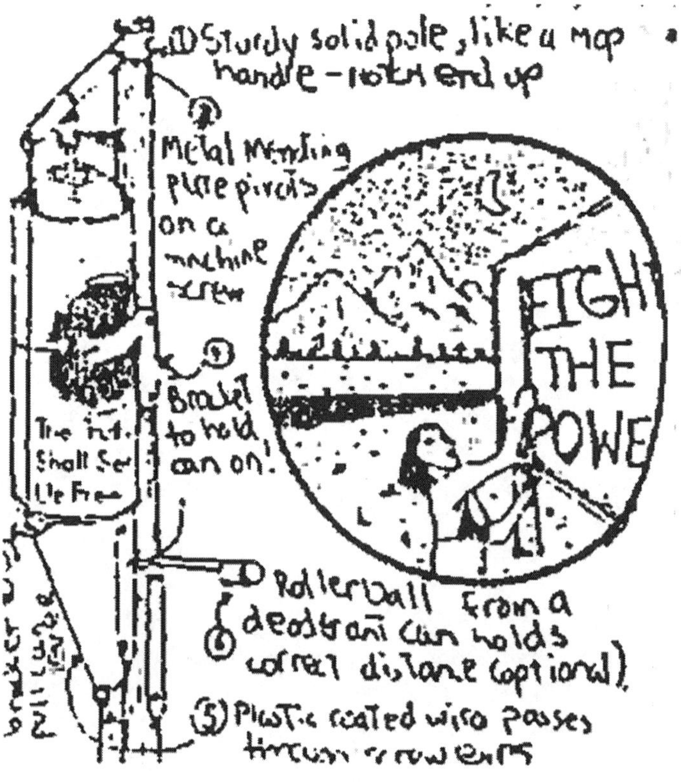

intended use. However, this diagram also depends on the reader's ability to extrapolate from examples and adapt to changing contexts. It limits itself to generalities, referencing a "sturdy solid pole, like a mop handle—notch end up" and references optional additions such as a "Rollerball from a deodorant can" to maintain a consistent distance between the rig and the chosen canvas. These components are named, but their dimensions, providence, and process of assembly are not specified. While in no way comprehensive, this diagram directs the reader's attention to a new and accessible tool, inviting the reader's further participation.

In both cases, the results of these instructions are not overdetermined, and the instruction itself is notably incomplete. In *BIKINI KILL*, there is a tacit acknowledgment of the impossibility of giving a complete account of the proposed activity. Instead, readers are expected to apply their own perspective and understanding to the task at hand. Its value lies in its provocation, suggesting new previously unconsidered courses of action and demonstrating their feasibility.

Embrace of Personal Experience as Knowledge Production

It is easy enough to imagine claims to authoritative knowledge arising from institutional accreditation, precedent, or access to information not readily available to society at large. In the context of technical communication, one could point to any number of installation or assembly instructions that demonstrate the force of these claims. They are identified through their association with the authoring institutions, and they are trusted because they are a familiar genre. They promise access to accurate, specific, and purposeful knowledge that would otherwise be difficult to find. These claims are persuasive in many contexts, but they also depend on institutional structures for their existence. Recognition, precedent, and specialization all require that significant resources and capacity be brought to bear in order to have their desired effects.

These institutional structures and the forms of authority they produce are not available to the creators of *BIKINI KILL*. In addition to the zine's DIY method of composition, the band was only recently formed and the zine produced in connection with its first tours in 1991. The relative fame of the riot grrrl movement had yet to manifest, so even if a wide range of readers were desired, the zine could not have counted on the band's notoriety for its success. Additionally, the medium's predominantly ad hoc methods of circulation through shows and informal networks of mailing addresses ensured that the zines' authors could not make a precise count of readership.

While the knowledge presented in this zine does not draw its value from conventional sources of authority, *BIKINI KILL* nonetheless expresses some notion of expertise as a part of its pedagogical function. The author(s) of *BIKINI KILL* insist that they "are not like this due to any weird gene formation or luck or trick. We are how we are from working together with our eyes open and having experience and getting help from out [*sic*] Moms and friends" (*BIKINI KILL: A COLOR AND ACTIVITY BOOK*, 1991, p. 1). While the authors claim some degree of distinction, admitting that they are in fact "like this" in contrast to an imagined norm, they describe this quality as a consequence of shared struggle and experience—a source of expertise accessible to their audience.

This emphasis on personal experience as a source of expertise is maintained throughout the content and arrangement of *BIKINI KILL*, resulting in the frequent use of evocative examples as pedagogical tools. Take for example, the "REVOLUTION girl STYLE now" section of the first issue of *BIKINI KILL*. In this section, the zine makes a series of instructions such as "ENCOURAGE IN THE FACE OF INSECURITY" and "MAKE PORNOGRAPHY that includes more than just hetero sex" (*BIKINI KILL: A COLOR AND ACTIVITY BOOK*, p. 11). Each instruction is accompanied by a short paragraph expanding on the potential that each instruction has for transforming this stubbornly patriarchal world. These potential effects range from forming new individual relationships based around shared creative endeavors and mutual education to composing and distributing fliers that address street harassment and other similar issues faced by the zine's readers. However, while *BIKINI KILL* encourages its readers to take the content seriously, it makes no claim to function as a totalizing set of instructions or to offer a stable end state for its audience. Instead, *BIKINI KILL* continues its pattern of deploying small series of evocative examples from which readers are expected to extrapolate and expand their understanding. By maintaining its focus on personal experience as the site of knowledge production, the zine rejects any universalizing program of social and political development while affirming the potential and necessity for social and political transformation.

Rejection of Cultural and Professional Norms

The study of technical communication, particularly in the context of the United States, has a long and ongoing commitment to the legitimate institutions of research, development, and production. Corporations, governmental programs, and professional organizations all maintain public-facing identities. Studies of TTC, however, have less of a commitment to these institutions and the conventions that accompany them. Indeed, it is in part the rejection of convention that results in much of TTC's distinct and complex character. Take, for instance, Miles Kimball's (2009) discussion of the DIY Locost car-building community, a group of predominantly "middle-class, middle-aged white men" that aim to build their own sports cars as an escape from the woes of a postindustrial society (p. 82). These enthusiasts collaborate with one another as part of a shared goal that rejects conventional narratives of class and responsibility, though they each realize this goal in an individual capacity. For its part, *BIKINI KILL* takes a double stand against the norms that structure the world. In the first place, this zine participates in the broad countercultural movement of its time, one organized against what was considered to be a consumerist culture empty of all vitality, dynamism, and authenticity. However, *BIKINI KILL* also makes a second movement, rejecting the misogyny endemic to the countercultural punk-music scene itself. There is no claim to innate exceptionalism on behalf of its authors or audience. Instead, the zine demonstrates an understanding of political struggle that depends on shared experiences and community practices of reflection and accountability.

Even within the communities of musicians and punks where the zines circulated, *BIKINI KILL*'s unwavering commitment to identifying and combating sexism guaranteed that it would not hold an uncontested position of legitimacy. These zines critiqued the contemporary punk-music scene and the ongoing presence of sexism within these supposedly liberatory spaces. Writing in response to a submitted letter, Bikini Kill's Kathleen Hanna notes: "Okay boys, the gig is up. Now that i got this letter i know it is possible for a boy to be truly cool so there is no excuse for being lamo jerks anymore. Okay?" (*BIKINI KILL GirlPower*, 1991, p. 20). The subjects of this passage's criticism are all members of the punk-music community, addressed as potential readers and participants in the imagined dialogue. This ongoing intra-community aspect of *BIKINI KILL*'s critique resulted in harsh and violent reaction. In a later 2017 interview, Hanna recounts how "people were throwing chains at our heads—people hated us—and it was really, really hard to be in that band" (Burbank, 2015). Escape on the level of the individual is not possible in this case. The community itself must adapt.

Of course, this is not to say that these zines or this band stand as unimpeachable models of cultural rebellion and communal liberation. The constraints of the medium, the context in which *BIKINI KILL* was composed, and the authors' individual inclinations speak to limits of scope, purpose, and enduring relevance. For example, issues of race and sexual orientation are discussed with comparative infrequency. Additionally, the zine's focus remains largely fixed on the United States and the countercultural punk-music communities therein.

These areas of emphasis grew and shifted with time as the broader riot grrrl movement continued to spread. The now well known "Riot Grrrl Manifesto," published in *BIKINI KILL GirlPower* (1991), gained an international audience, and the band inspired other punk feminist music groups and communities (Marcus, 2010). While *BIKINI KILL* rejected both broad cultural norms and specific conventions within the Pacific Northwest punk-music scene, it nonetheless found widespread acclaim and rhetorical reach.

Implications for the Histories of TTC

In its own right, *BIKINI KILL* demonstrates how feminist historiography can help to contextualize individual instantiations of TTC and the wider trends they participate within. This context includes the conditions that surround any given example's production, circulation, and imagined use, and such an attention to context is common in all useful historical scholarship. Feminist historiography builds upon this practice, examining the role of these specific examples within broader discussions of the field's past and ongoing development. It asks us to imagine previously unconsidered histories and to examine the processes by which these histories are granted legitimacy.

Feminist historiography's embrace of the partial, partisan, and uncooperative characteristics of TTC also equips scholars well for confronting the challenges that come with studying technical communication outside of the institutions that so often characterize it. As a methodology, feminist historiography encourages a reinterpretation of conventional wisdom, a reevaluation of the worth of an academic canon, and an attention to otherwise ignored sources of knowledge. As a body of rhetorical scholarship, it suggests some of the potential outcomes that may come from folding examples of TTC into existing histories of the field, and it gestures toward the value of understanding TTC as a tradition of rhetorical practice with histories, characteristics, and a future all its own.

Bringing feminist historiography's emphasis on the political dimensions of history into our scholarship may also help offer a perspective for scholars who aim to answer recent calls for further discussion of the political and ethical dimensions of technical communication in general and TTC in particular (Walton et al., 2019; Colton et al., 2017). Feminist historiography attends to the long-term effects of such processes—asking scholars to frame rhetorical acts in the context of broader political struggles.

Take, for example, the relationship between contemporary technology companies and their increasing use of user-generated technical communication (Swarts, 2018). Within this trend, one can identify certain characteristics associated with TTC—such as a decentralized model of expertise built around individual hobbyists and tinkerers. However, the maintenance of formal and legal hierarchies over said platforms works to ensure that the results of these processes do not challenge or disrupt these companies' existing strategic goals.

New forms of TTC are created out of this sort of friction between communities of users and extant institutions. *BIKINI KILL* is one such historical example, but TTC continues to be produced in contemporary contexts as well. New resources and constraints inform the process of creating such documents: widespread access to design software, online communities positioned against dominant cultural norms, the consistent potential for monetization, corporate ownership of social media platforms, aggressive enforcement of intellectual property law. Among many others, these factors inform what new examples of TTC are being produced today. While the context that shaped *BIKINI KILL* and other zines of its time has been irrevocably changed during the ensuing decades, new texts that seek a similar blend of instruction, agitation, and expression can come to be in this new context.

Like any recognizable and enduring rhetorical practice, TTC is in no small part defined by its histories. Its inspirations, texts, habits, and effects shape how ambitious, creative, and desperate practitioners create new forms of knowledge away from the watchful eyes of the powers that be. The impact of these histories is felt in the present, whether it is recognized as such or not, and technical communication would benefit from the perspective they offer.

Notes

1. *Technical Communication After the Social Justice Turn* is the most contemporary of the contributions cited here. In it, Walton, Moore, and Jones (2019) draw from explicitly Black feminist theories to articulate a purpose for social-justice

research in technical communication as an investigation of "how communication broadly defined can amplify the agency of oppressed people" (p. 50).

2. One notable exception to this trend tends to be fan-submitted letters. These are often identified as such and commented upon by the authors, perhaps due to the importance of such engagement to zine making as a whole.

3. This diagram, while not captioned in *BIKINI KILL*, also appears on the back cover of *The Art & Science of Billboard Improvement*. That version of the diagram lacks accompanying annotations and is credited to the environmentalist group Earth First!

References

Biesecker, B. (1992). Coming to terms with recent attempts to write women into the history of rhetoric. *Philosophy & Rhetoric, 25*(2), 140–161. https://www.jstor.org/stable/40237715

BIKINI KILL: A COLOR AND ACTIVITY BOOK [zine] (1991).

BIKINI KILL GirlPower [zine] (1991).

Billboard Liberation Front and Friends. (2000). *The art & science of billboard improvement* (Second edition). Los Cabrones.

Burbank, M. (2015, April 22). *Rebel girl, redux*. The Portland Mercury. https://www.portlandmercury.com/portland/rebel-girl/Content?oid=15463567

Campbell, K. K. (1993). Biesecker cannot speak for her either. *Philosophy & Rhetoric, 26*(2), 140–161. https://www.jstor.org/stable/40237761

Colton, J. S., Holmes, S., & Walwema, J. (2017). From NoobGuides to #OpKKK: Ethics of anonymous' tactical technical communication. *Technical Communication Quarterly, 26*(1), 59–75. https://doi.org/10.1080/10572252.2016.1257743

de Certeau, M. (2011). *Practice of everyday life* (S. F. Rendall, Trans.). University of California Press.

Duncombe, S. (2017). *Notes from underground: Zines & the politics of alternative culture* (3rd edition). Microcosm.

Durack, K. T. (2004). Gender, technology, and the history of technical communication. In J. Johnson-Eilola & S. A. Selber (Eds.), *Central works in technical communication* (pp. 35–43). Oxford University Press.

Edenfield, A. C., Holmes, S., & Colton, J. S. (2019). Queering tactical technical communication: DIY HRT. *Technical Communication Quarterly, 28*(3), 177–191. https://doi.org/10.1080/10572252.2019.1607906

Enoch, J. (2013). Releasing hold: Feminist historiography without the tradition. In M. Ballif (Ed.), *Theorizing histories of rhetoric* (pp. 58–73). Southern Illinois University Press.

Glenn, C. (1997). *Rhetoric retold: Regendering the tradition from antiquity through the renaissance.* Southern Illinois University Press.

Johnson-Eilola, J., & S. A. Selber. (Eds.), *Central works in technical communication* (pp. 146–159). Oxford University Press.

Kimball, M. A. (2009). Cars, culture, and tactical technical communication. *Technical Communication Quarterly, 15*(1), 67–86. https://doi.org/10.1207/s15427625tcq1501_6

Lay, M. M. (2004). Feminist theory and the redefinition of technical communication. In J. Johnson-Eilola & S. A. Selber (Eds.), *Central works in technical communication* (pp. 146–159). Oxford University Press.

Logan, S. W. (1999). *"We are coming": The persuasive discourse of nineteenth-century black women.* Southern Illinois University Press.

Marcus, S. (2010). *Girls to the front: The true story of the riot grrrl revolution.* Harper Perennial.

Sarat-St. Peter, H. A. (2017). "Make a bomb in the kitchen of your mom": Jihadist tactical technical communication and the everyday practice of cooking. *Technical Communication Quarterly, 26*(1), 76–91. https://doi.org/10.1080/10572252.2016.1275862

Sarat-St. Peter, H. A., & St. Peter, A. L. (2020). "Figure 4, Peyote": Comics and graphic narrative in anarchist cookbooks, 1971–present. *Technical Communication Quarterly, 29*(3), 255–270. https://doi.org/10.1080/10572252.2020.1768293

Swarts, J. (2018). *Wicked, incomplete, and uncertain: User support in the wild and the role of technical communication.* Utah State University Press.

VanHaitsma, P. (2016). Gossip as rhetorical methodology for queer and feminist historiography. *Rhetoric Review, 35*(2), 135–147. https://doi.org/10.1080/07350198.2016.1142845

Walton, R., Moore, K. R., & and Jones, N. N. (2019). *Technical communication after the social justice turn.* Routledge.

Chapter Twelve

Our Bodies, Ourselves

A Case for Metical Technical Communication

Kevin Van Winkle

It has been roughly 15 years since Kimball (2006) introduced his theoretical framework for tactical technical communication (TTC). A review of the literature that followed reveals that researchers have used Kimball's work to investigate a variety of topics and contexts. It also shows a trend of researchers using Kimball's analytical lens to examine user-producers who create, share, and use tactical information as it relates to the human body—how to heal it, how to change it, and how to reclaim control of it.

In this chapter, I review scholarship that investigates the intersection of TTC and the body. I go on to argue this trend warrants a reconsideration of the ancient Greek concept *metis* and its relationship to tactics. Typically defined as "cunning intelligence," metis is linked with tactics in the work of both de Certeau (1984) and Kimball (2006); however, I reason, while metis is very similar in meaning to *tactics*, it is ultimately distinguishable from it by its corporeal nature. In other words, metis is additive, containing all of the aspects and features of tactics—outsider status, *bricolage*, *la perruque*—but with a necessary embodiment of those tactics.

I then go on to demonstrate how a concept like metical technical communication can work analogously to Kimball's original analytical framework for TTC by examining the case of *Our Bodies, Ourselves*. In doing so, I expand upon Kimball's concept, while also fulfilling his objective to broaden the scope of technical communication. I then go on to justify how metical technical communication can be used to subvert dominant cultural and organizational strategies designed to control and manipulate marginalized bodies of all types.

Strategies, Tactics, and Technical Communication

Kimball (2006) characterized strategic technical communication as the type of technical communication that comes from official institutions. The manufacturer's instructions, the information on a company's website, and city maps are all examples of strategic technical communication. Conversely, TTC is *extra-institutional.* It's created by user-producers (Johnson, 1998), people who bring their own experiential knowledge to bear upon how they use and reuse strategically produced texts or products. These user-producers are tacticians. Or, relative to technical communication, they are tactical technical communicators.

Kimball's (2006) conception of TTC augments tactics originally identified by de Certeau (1984) while also adding some unique ones specific to technical communication. For instance, de Certeau's (1984) bricolage originally described a person making something for their own use by creatively combining what's available to them. Paraphrased as "making do," the tactical home chef who whips up something delicious from the ingredients they have on-hand in their kitchens exemplifies bricolage. Bricolage in TTC is a user-producer who pulls information from multiple sources—the manufacturer's website, the insert that came with the product, their own experiences—to make sense of and potentially solve some sort of technical problem.

A memorable example of bricolage in TTC comes from Kimball himself. In a presentation on the topic, Kimball showed a picture of a skyscraper under construction with a large steel beam jutting out. On this beam was a chalk line intersecting perpendicularly. On the right, above an arrow pointing to the line, it read, "Cut here." No doubt, as Kimball noted, there were blueprints and plans, experts and inspectors, tools and instruments that all comprised and communicated the official strategy for cutting this beam. Yet one tactically minded steel worker practicing bricolage was able to produce a simpler method—a shortcut—by making do with what was available to them several stories in the air.

La perruque "is the worker's own work disguised as work for his employer" (de Certeau, 1984, p. 25). It "involves appropriating time or surplus material at work to personal uses" (Kimball, 2006, p. 72). An administrative assistant writing a personal letter on company time or an artisan borrowing a tool from their employer to work on a project at home are straightforward examples of *la perruque*. If, instead of a personal letter, it was instructions for car repair or if the tool were AutoCAD purchased and licensed to the company, then it could be considered *la perruque* in TTC.

Kimball (2006) also explained that experiential know-how is a common technique used in TTC. Tactical narratives enable the communicator to demonstrate "how they did it," as opposed to institutional strategies that dictate "this is how it should be done." These narratives exhibit both a local and communal function. By telling the story of "jailbreaking" their iPhone, the tactical technical communicator tells the story of their own tricks and techniques at the same time they tell a meta story of how practically any person can use the same tactics to adapt and subvert the rules enforced upon them by a capitalistic society and a powerful corporation that goes to great lengths to strategically make their products difficult—impossible for most really—to repair or upgrade, as Apple does.

Kimball (2017) also added another tactic: radical sharing. He acknowledged that de Certeau could not have envisioned the affordances in speed and scope engendered by the internet. At the same time, Kimball argued that the affordances enabled by the internet to increase the range of distribution for tactics, while simultaneously decreasing the time it takes to do so, is so profound that it should be considered a tactic in and of itself. The internet's efficiency makes radical sharing "profoundly connected to technical communication" because technical communication—as well as TTC—is "about making things happen" (Kimball, 2017, p. 4). In other words, what's "radical" about the sharing of TTC facilitated by the internet is that a greater number of people can share it at faster speeds, which means more things can happen—and typically do.

TTC and Bodies

While many scholars have answered Kimball's (2006) call to broaden our view of technical communication and examine the various ways extra-institutional documents are created and shared by tactical technical communicators (Ding,

2009; Van Ittersum, 2014; Colton et al., 2017; Sarat-St. Peter, 2017), a closer review of the literature reveals a significant portion of scholarship focused on body-related artifacts and practices. Indeed, many scholars have studied the ways marginalized user-producers have used TTC to examine, heal, change, or otherwise control their bodies and by doing so reclaim their bodily autonomy from the powerful strategic forces that exercise power over it.

A prominent entry in the area of embodied TTC scholarship is Hallenbeck's (2012) analysis of user-created instructional materials for women invested in the 1890s bicycling trend. She showed how these do-it-yourself manuals enabled women to become physically empowered and, in some ways, overcome the limitations prescribed to their strategically gendered, and therefore marginalized, bodies. She explained, "We can see their writing as an effort to transform the gender order by amplifying, though their texts, the rhetorical impact of otherwise unrecorded embodied practices aboard the bicycle" (p. 292).

A number of scholars have examined the role TTC plays in pushing back against strategies that relegate bodies. For instance, McCaughey (2021) examined how user-producers of an "exclusive pumping" site shared their personal stories of breastfeeding, or being unable to, in order to demonstrate how "tactical mothers" eschew the judgment of pediatric healthcare professionals and others who malign the practice of exclusive pumping. Relatedly, Seigel (2013) explored the issue of disembodiment, motherhood, and TTC in the rhetoric of pregnancy. Edenfield and colleagues (2019a, 2019b, 2019c) have done significant work "queering" TTCs to show how they support "a wide range of queer spaces, bodies and communicative practices" (2019a, p. 178). And similar to the main theme of this chapter, Yusuf and Schioppa (2022) asserted that YouTube instructional videos on Black hair care represent a form of TTC situated at the intersection of practical expertise and social justice advocacy.

Other scholars have investigated TTC as it relates to at-risk bodies. Bellwoar (2012), for example, explored patients' tactical use and (re)production of strategic texts created by official medical institutions to make sense of their strategically diagnosed bodily ailments. Working within a larger argument for mitigating the irrelevance of the whole body that some healthcare professionals have traditionally exhibited, Holladay (2017) examined online forums and the tactics created and shared there by people who have been formally diagnosed with mental "conditions," identifying embodied narrative as the primary tactic used to subvert institutional technical communication that predominantly situates patients as members of "(*disembodied*) categories of the DSM" (p. 18). Likewise, Bivens and colleagues (2018) posited that the adjustments people make to their bodies and to the technological tools they use to track and treat type 1 diabetes are a type of TTC. And Bishop and colleagues (2022) discussed how embodied risk communication during the COVID-19 pandemic required both strategic (conscious, institutional) and tactical (oppositional, individual) elements.

Metis Defined

Although important in ancient Greek culture, the "cunning intelligence" labeled metis was never the subject of any formal treatment by the Greeks. Creating such a treatise seemed inherently incongruent because metis was considered deeply intertwined with *kairos* and too heavily reliant on context; it would be like trying to teach street smarts through a textbook. Instead, metis was conveyed and understood through myth and stories where the characters exemplified it. Odysseus, for instance, was given the epithet *polumetis* (many + metis) because of the cunning intelligence he displayed throughout the *Odyssey*. Consequently, it remains a mostly understudied concept to this day. Indeed, those few who have sought to understand metis often rely heavily on a single text: Detienne and Vernant's (1978) *Cunning Intelligence in Greek Culture and Society*. In this book, the authors provided what has become the foundational definition of metis: "a type of intelligence and of thought, a way of knowing; it implies a complex but very coherent body of mental attitudes and intellectual behavior which combine flair, wisdom, forethought, subtlety of mind, deception, resourcefulness, vigilance, opportunism, various skills, and experience acquired over the years" (p. 3).

Notwithstanding the unique significance of their work and the practicality of equating metis with "cunning intelligence," Detienne and Vernant (1978) have been criticized for their elision of the inherent role the body plays in performing metis. They built their definition of metis from the same stories and myths the Greeks used to portray it; yet, as others have pointed out, they often failed to highlight that the characters who displayed metis did so via their bodies. As Hawhee (2005) explained, "The particular styles of becoming illustrated by these figures of metis thus underscore the corporeality of metis: as a kind of intelligence, metis cannot be thought separate from bodily state" (p. 57).

A closer look at the stories of metis supports this claim. Oppian theogony, for instance, explained metis's origin story thusly: Zeus's first wife, the Titaness Metis, was more wise and cunning than all the other gods and mortals. Troubled by predictions that their future children would eventually grow to overthrow him, Zeus devoured Metis—literally embodying her—while she was pregnant with their first child, Athena, and thereby transmogrified into metis itself/himself. In other areas of Oppian myth, metis is demonstrated specifically by cunning animals who overcome their perceived physical limitations to win out over their ostensibly more physically powerful opponents. When the crafty octopus thinks to shove its tentacle down the predator eel's throat, it has used metis. And it is the fox who plays dead and disguises its body as a corpse so that it can seize upon any birds curious enough to investigate that demonstrates metis.

Like Detienne and Vernant (1978), from whose work he builds his definition, de Certeau (1984) used the shorthand "cunning intelligence" to define metis and ultimately link it to his concept of tactics. Also like them, his descriptions of metis allude to its necessary corporeality but never make it explicit. He explained that metis "is close to everyday tactics through its 'sleights of hand, its cleverness and its stratagems' " (p. 81). Although ostensibly a figure of speech, one also borrowed from Detienne and Vernant, the phrase *sleights of hand* suggests corporeality as a characteristic of metis. Furthermore, the linkage of *cleverness* and *stratagems* with the *hand* indicates that de Certeau viewed tactics as *close* to metis but with at least one difference being a corporeal element. Understanding de Certeau's definition of metis this way makes for a useful categorization for the current argument as it shows that metis is not distinctly different from tactics but rather very similar to it with the additional criteria of embodiment.

If the idea that metis is additive to tactics—or at least a specific type of embodied tactics—is permissible, then it follows that a concept like metical technical communication could be considered a kind of TTC and therefore a potential supplement to Kimball's original framework. Metical technical communication maintains all the same features Kimball (2006, 2017) assigned TTC—outsider status, use of narrative, bricolage, *la perruque*, radical sharing—but it is used in the service of understanding and reclaiming the human body from the strategic controls exercised upon it. To put it another way, metical technical communication is the "weaker" marginalized individual making do with technical communication that operates outside of and against the strategic systems of stronger institutions, with the body centralized in this struggle.

Our Bodies, Ourselves as Metical Technical Communication

The 1971 first edition of *Our Bodies, Ourselves* and the story of its creation display all the hallmarks of TTC as theorized by Kimball (2006). It was created by outsiders using borrowed time and tools to compile information from various sources of personal experience and professional documentation and then shared in radical ways and to radical effect. At the same time, all of these tactics were employed so that women could better understand their bodies. And through that understanding, they could take back control of their bodies from a society that had marginalized and oppressed them. Combined, these elements make *Our Bodies, Ourselves* a useful archetypical artifact of metical technical communication.

The genesis of *Our Bodies, Ourselves* came from a women's health workshop conducted at a liberation conference held in 1969 at Emmanuel College in Boston. For the women who attended, the frank discussions and exchange of information regarding their health and bodies were revelatory and empowering. Subsequently, a smaller group of these women calling themselves the Boston Women's Health Book Collective decided there was much more that needed to be said and shared about women's health and bodies. Together, they decided on important topics, such as sexuality, anatomy, birth control, abortion, pregnancy, medical laws, and organizing for change, and then each chose one to research further, write about, and return to share with the group. Eventually, the group had a corpus of texts that covered a wide range of issues and topics. They then shared and used these texts to conduct workshops with other women, who then used and shared them in other workshops and with other women, so on and so forth.

In light of this growth, the group decided to revise their papers and collect them together into a single manuscript that could be mimeographed and shared more easily. They titled the document *Women and Their Bodies* and published 5,000 copies. This first run sold out very quickly, creating an urgent need for a second. While the content of the book remained unchanged between printings, the group decided to retitle it *Our Bodies, Ourselves.* The title change had the express purpose of encouraging women to reclaim ownership of their own bodies.

While it would be difficult to quantify the number of women who read and used these first printings of *Our Bodies, Ourselves* (if nothing else for the reason that at the time it was an illicit document many women felt compelled to hide and keep secret), its cultural impact and subsequent growth suggest it had to have been significant. Following its initial publication, *Our*

Bodies, Ourselves has been revised and expanded nine times. It has sold over 4 million copies and been translated into 33 languages. It is a canonical text of feminism and has been designated by the Library of Congress as one of the Books That Shaped America. Suffice it to say, it seems that even without exact measurement, the impact of this book was and is massive.

Although the workshops created around *Our Bodies, Ourselves* provided instruction for a broad range of body-related issues, the book itself did not give many straightforward directions for how to do things. Instead, it provided women with knowledge to help them understand their bodies, which was the first step to pushing back against the patriarchal institutions that had controlled them through strategies of enforced ignorance and shame. The authors explained:

> As women, knowledge of our reproductive organs is vital to overcome objectification. We have been ignorant of how our bodies function and this enables males, particularly professionals, to play upon us for money and experiments, and to intimidate us in doctors' office[s] and clinics of every kind. Once we have some basic information about how our bodies work by talking and learning together and spreading the correct information, we need not be at the total mercy of men who are telling us what we feel when we don't or what we don't feel when we do. (p. 10)

The evidence for *Our Bodies, Ourselves* as metical technical communication is clear in its very first sentence: "One year ago, a group of us who were then in women's liberation . . . got together to work on a laywoman's course on health, women and our bodies" (Boston Women's Health Collective, 1971, p. 3). The authors' identification as members of the women's liberation movement, their use of the term *laywoman*, and the topic of women's bodies all demonstrate the metical nature of the text. The book and its authors are positioned outside and opposed to the larger strategic apparatuses of a patriarchal medical establishment and its specialists, who had failed to include women in their research and practices adequately. This sentiment is reiterated in the authors' statement of exigence: "We discovered there were no 'good doctors' and we had to learn for ourselves" (p. 3).

Also evident in the first sentence of the book is that personal narratives are the predominant mode of transfer and explanation. Time and context are introduced. So, too, are the heroes—the women who wrote it. The heroes' quest is established: to understand bodies and share that understanding with others. And this is all communicated with subjective first-person narration. Kimball (2006) called this type of storytelling "technological narrative" for the car hobbyists who shared their technical know-how this way. For these hobbyists, their stories became a way to understand and potentially subvert "institutional practices embodied by technological artifacts" (p. 75). It is similar in the case of *Our Bodies, Ourselves* but with a major distinction: the women user-producers of it used and shared metical narratives to understand and subvert the "institutional practices" they themselves embodied.

In addition to sharing metical narratives, the authors of *Our Bodies, Ourselves* encouraged their readers to become bricoleurs and to collect information wherever they could find it, just as they themselves had done to construct the book. To craft *Our Bodies, Ourselves*, the authors located what useful research they could from academic institutions, helpful medical professionals, and other strategic sources, such as the companies that made products specifically for women's bodies (e.g., birth control pills and tampons). They make their own bricolage evident throughout. There are citations indicating the scholarly sources where they found certain information. They point to specific books they used, such as Lennart Nilsson's *A Child Is Born: The Drama of Life Before Birth in Unprecedented Photographs; A Practical Guide for the Expectant Mother* and the McGill University Students' Society's *Birth Control Handbook*, and encouraged readers to access for themselves. The authors also advocated for finding "good" medically trained professionals and visiting allied groups like Planned Parenthood and the Women's Abortion Project to get useful information, as they had done. And, although they warned readers about getting the information they needed from "capitalist organizations, pushing their own products for profit," (Boston Women's Health Collective, p. 5), they even recommended making do with the information that can be garnered from strategically aligned companies, such as information provided by Tampax's educational department.

More evidence of *Our Bodies, Ourselves* as metical technical communication comes from the authors' use of *la perruque*. If, as de Certeau (1984) explained, *la perruque* is doing one's own, personal work on "company" time, then it's difficult to find any aspect of the construction and subsequent publishing and sharing of *Our Bodies, Ourselves* that isn't an example of it. If the larger patriarchal society and culture at the time were the entities that exerted strategic control over women by denying them an understanding of their bodies and health, then for the women who created *Our Bodies, Ourselves*, every second spent attending meetings and workshops, or researching and writing, or sharing and teaching information about women's bodies would necessarily have been *la perruque*. To put it another way, for prisoners, there is no such thing as free time.

Initially, it may seem that *Our Bodies, Ourselves* could not have employed another essential tactic of TTC: radical sharing. According to Kimball (2017), radical sharing is partly a condition of the internet and the "sharing of tactics with people the world over at great speed and with great effect" (p. 4) that it allows for. Given the time of its creation, *Our Bodies, Ourselves* obviously could not have benefitted from internet-enabled fast-and-wide distribution; however, this seemingly rational explanation is belied by a reconsideration of the word *radical*. In one sense, the descriptor is intended to characterize sharing that is done between a large number of people at a fast speed. These sorts of quantitative measures of time and scope are qualifiable, though. Surely, for many people riding in a car at 40 mph for the first time must have felt extremely fast. Likewise, a bit of gossip shared with the entire community seems extreme in scope, even if that community is only a 100 people. Relatively speaking, then, going from a single conference session to the formation of a group sharing papers in weekly meetings to collecting and publishing these papers in a book that sold out so quickly it needed a subsequent printing in months, ultimately going on to be shared with countless other women across the nation and then the world can be considered a prime example of radical sharing.

The other—and more meaningful—way *Our Bodies, Ourselves* can be viewed as having participated in the tactic of radical sharing is also substantiated semantically. *Radical* also denotes momentous changes in political views, practices, or policies. There is evidence to show that this sense of the word was also intended by Kimball (2017). In his explanation of radical sharing as a type of tactic, he explained, "Radical sharing is profoundly connected to technical communication. It's not about what happened, but about making things happen" (p. 4). And this in part is what makes TTC so powerful. Certainly, then, if making things happen and being powerful are measures for radical, then *Our Bodies, Ourselves* meets the criteria.

Conclusion

One of Kimball's (2006) stated intentions was to expand the scope of technical communication and to advocate for examining meaningful artifacts of technical communication wherever we find them. A theoretical framework of metical technical communication is an attempt to do the same.

It is intended to describe a type of TTC that makes the role of the body prominent and underscores the preexisting inclination of scholars to view TTC as an embodied practice. Whereas Kimball advanced his theoretical framework for TTC as one with which we can investigate the intersections of "technology, discourse, and people's lives" (p. 84), metical technical communication encourages us to examine the intersections of technology, discourse, people's lives, and bodies. In both cases, there is the potential to remake not just our discipline but the world itself.

Some might consider this too grand a claim. Admittedly, Kimball's (2006) initial example of tactical technical communicators seems rather un-radical: car hobbyists. Nonetheless, rather than limit the applicability of TTC to a taxonomy of the DIY ethic so pervasive in our culture, it would be best for us to consider the very large and very real political ramifications of the strategic forces that shape our lives and examine how tactical and metical technical communication enable us to understand and potentially subvert them. Certainly, there seems like no better time than the present for doing so.

During the course of writing this chapter, we have moved into a post-*Roe* America. And what I once wrote as potential had to be revised into past tense. Forced birth, deadly medical complications going untreated, and the imprisonment of pregnant women to prevent them from exercising their bodily autonomy have already happened. And it appears such outrages and horrors will continue to happen for the unforeseeable future. The strategies the authors of *Our Bodies, Ourselves* and all those who read and used it fought so hard to overthrow through metical technical communication are being reestablished at a frightening speed. Furthermore, we are reminded daily of the way other marginalized bodies—Black, Indigenous, and people of color's bodies, disabled bodies, trans and queer bodies—suffer from the strategies imposed upon them by judicial, penal, and medical institutions. All of this suggests that the power of TTC and metical technical communication cannot be understated. Through them, the capacity to analyze, understand, and advocate for undermining the strategic controls exerted upon us in all ways is not just possible but critically important and urgent.

References

Bellwoar, H. (2012). Everyday matters: Reception and use as productive design of health-related texts. *Technical Communication Quarterly, 21*(4), 325–345. https://doi.org/10.1080/10572252.2012.702533

Bishop, T., Capan, E., Larsen, B., Preston, R., & Sparby, E. M. (2022). Tactical risk communication: Observations from teaching and learning about crisis communication during COVID-19. *Technical Communication Quarterly, 31*(2), 175–189. https://doi.org/10.1080/10572252.2021.2008509

Bivens, K. M., Arduser, L., Welhausen, C. A., & Faris, J. (2018). A multisensory literacy approach to biomedical healthcare technologies: Aural, tactile, and visual layered health literacies. *Kairos: A Journal of Rhetoric, Technology, and Pedagogy, 22*(2). Online Publication. http://hdl.handle.net/2346/74545

Boston Women's Health Collective. (1971). *Our bodies, ourselves.* New England Free Press.

Colton, J. S., Holmes, S., & Walwema, J. (2017). From NoobGuides to# OpKKK: Ethics of anonymous' tactical technical communication. *Technical Communication Quarterly, 26*(1), 59–75. https://doi.org/10.1080/10572252.2016.1257743

de Certeau, M. (1984). *The practice of everyday life.* University of California Press.

Detienne, M. & Vernant, J. (1978). *Cunning intelligence in Greek culture and society* (J. Llyod, Trans). University of California Press.

Ding, H. (2009). Rhetorics of alternative media in an emerging epidemic: SARS, censorship, and extra-institutional risk communication. *Technical Communication Quarterly, 18*(4), 327–350.

Edenfield, A. C., Colton, J. S., & Holmes, S. (2019a). Always already geopolitical: Trans health care and global tactical technical communication. *Journal of Technical Writing and Communication, 49*(4), 433–457. https://doi.org/10.1080/10572250903149548

Edenfield, A. C., Colton, J. S., & Holmes, S. (2019b). Healthcare and global tactical technical communication in online trans-DIY forums. *Journal of Technical Writing and Communication*, 433–457. https://doi.org/10.1177/0047281619871211

Edenfield, A. C., Holmes, S., & Colton, J. S. (2019c). Queering tactical technical communication: DIY HRT. *Technical Communication Quarterly, 28*(3), 177–191. https://doi.org/10.1080/10572252.2019.1607906

Hallenbeck, S. (2012). User agency, technical communication, and the 19th-century woman bicyclist. *Technical Communication Quarterly, 21*(4), 290–306. https://doi.org/10.1080/10572252.2012.686846

Hawhee, D. (2005). *Bodily arts.* University of Texas Press.

Holladay, D. (2017). Classified conversations: Psychiatry and tactical technical communication in online spaces. *Technical Communication Quarterly, 26*(1), 8–24. https://doi.org/10.1080/10572252.2016.1257744

Johnson, R. R. (1998). *User-centered technology.* State University of New York Press.

Kimball, M. A. (2006). Cars, culture, and tactical technical communication. *Technical Communication Quarterly, 15*(1), 67–86. https://doi.org/10.1207/s15427625tcq1501_6

Kimball, M. A. (2017). Tactical technical communication. *Technical Communication Quarterly, 26*(1), 1–7. https://doi.org/10.1080/10572252.2017.1259428

McCaughey, J. (2021). The rhetoric of online exclusive pumping communities: Tactical technical communication as eschewing judgment. *Technical Communication Quarterly, 30*(1), 34–47. https://doi.org/10.1080/10572252.2020.1823485

Sarat-St. Peter, H. A. (2017). "Make a bomb in the kitchen of your mom": Jihadist tactical technical communication and the everyday practice of cooking. *Technical Communication Quarterly, 26*(1), 76–91. https://doi.org/10.1080/10572252.2016.1275862

Seigel, M. (2013). *The rhetoric of pregnancy.* University of Chicago Press.

Van Ittersum, D. (2014). Craft and narrative in DIY instructions. *Technical Communication Quarterly, 23*(3), 227–246. https://doi.org/10.1080/10572252.2013.798466

Yusuf, M., & Schioppa, V. N. (2022). A technical hair piece: Metis, social justice and technical communication in Black hair care on YouTube. *Technical Communication Quarterly, 31*(3), 263–282. https://doi.org/10.1080/10572252.2022.2077454

Chapter Thirteen

Armed Propaganda and the Ethics of Horrorism

Tactical Communiques from the Weather Underground

Brad Lucas

Tactical technical communication (TTC) offers productive frameworks not only for contemporary social justice work but also for investigating historical antecedents of tactical practices. Central to TTC are Michel de Certeau's formulations about agency and resistance (1997), which he developed amidst the backdrop of sixties revolutionism (e.g., the May 1968 protests in *The Capture of Speech and Other Political Writings*). In this chapter, I offer a case study centered on this very era, tracing the complex TTC dynamics in the Weather Underground Organization that emerged from the movement organization Students for a Democratic Society (SDS) in 1969 and became a network of clandestine working "foco groups" we now categorize as terrorist cells. Through historicizing tactical texts, we can better understand the zeitgeist around TTC's origins and highlight the ethical tensions around individual and collective agency against the state, all exacerbated by extremism, ideological discipline, and the coordinated messaging campaigns that used violent spectacle for citizen engineering. Given that today's right-wing extremism echoes the dynamics of sixties left-wing militancy, we can benefit from seeing TTC as it manifests in public spectacles, underground cell communications, and the myriad ways a decentralized movement (un)ethically instructs and organizes individual agency and collective action.

The story of the Weather Underground reveals the TTC changes in a nationwide organization whose members lost faith in institutional reformation and turned instead to confrontational "direct-action" tactics of symbolic vandalism, now a globally familiar practice among contemporary Black Bloc groups. Rather than merely catalogue the manifold instances of TTC within Weather, however, I historicize key dynamics of TTC and take up the charge from Colton, Holmes, and Walwema (2017) to consider the ethical dimensions of TTC through Adriana Cavarero's (2011) framework of horrorism, with a specific focus on tactical practices designed to mobilize militant action against the state. Through Weather's use of public spectacle, management of individual subjectivity, and its headlong embrace of violence, we can trace the contours of a TTC that loses its care for individual and community vulnerability and, consequently, its ethical bearings: reflecting in its own tactical work the horrors of the institutional system it reviled yet ultimately changing its approach once facing the consequences of its own horrorist agenda.

Active Equality, Radical Sharing, and Alienation

Drawing on the work of Jacques Rancière, Colton and Holmes (2018) argue that social-justice researchers in technical communication typically and tacitly operate within "passive-equality" political-organization frameworks that rely on institutional, system, or state-sponsored solutions, whereas "active equality" aims to enable agency beyond those solutions: "[A]n active theory of social justice recognizes that equality is something that any individual, including professional technical communicators, can enact independent of a permissive institutional or governmental structure" (p. 12). Social justice is thus enacted whenever it "makes visible the equality of even one person whose voice has been suppressed and whose equality has been erased or ignored" (p. 13), and importantly it accounts for more than just an opposition to institutions. Active-equality social justice is, instead, a motivation against "the police order"—structures of the state, not law enforcement—which entails "the establishment of communicative and behavioral norms as they are invented, circulated, reaffirmed, and produced to be then distributed to how bodies are ordered by these norms" (p. 15). Active equality is a politics motivated without need for, or heed to, policy or the systems and mechanisms of the state to do its work: it is "a practice of constant verification to strive for . . . an active form of social justice that anyone can practice"

(p. 16). Working beyond the strategic aims and norms of the police order, the politics of active equality centers the individual, and individual agency, in the service of social justice.

Technical communication predicated on active equality, and navigating the police order, often relies on individuals working tactically within the strategic structures of dominant institutions or organizations. Anchoring de Certeau (1984)—with Kimball's (2006) formation of TTC—to the late-sixties zeitgeist is important for our larger cultural and historical orientation toward TTC, particularly on an ethical level wherein individual and collective dynamics are in tension. Kimball situates a 1969 Volkswagen Beetle repair manual in the counterculture and antitechnocracy spirit of the Sixties (and its legacies), noting how the manual revealed tensions about technology and the bureaucratic culture controlling it: "a tactical response to the strategies of military, corporate, and industrial technocracy (p. 77). The maneuvering of the systems-strategies of technocracy reflects not just a cultural narrative of resistance but points to those "contradictory desires" behind tactical motives: they embrace the appropriation of the technological while opposing—if not dismantling—the strategic hegemony of the technocratic. Such contradictions can reflect an ideological balancing act for TTC producers who consequently risk exposure on the ethical plane when the tactics incur, deploy, or otherwise leverage violent technologies in their aims against a violent technocracy. Tactics have to navigate the hegemony of the strategic order in the process of re-creating and repurposing technology, and all such processes will be received by observers—readers, bystanders, and other audiences—in myriad ways because of that hegemony.

Kimball's (2017) "radical sharing" accounts for the collective dimension of individual tactics, wherein instances of TTC are multiplied and made powerful as a catalyst for collective action, for "citizen engineering" (p. 4). Instructions for making bombs, preparations for a violent police confrontation, or survival tips for evading FBI capture emanate from individuals who advocate for citizen engineering against the state. When "the people engaged in radical sharing are actually radicals" (p. 6), as in the cases of Anonymous (Colton et al., 2017) and Jihadist organizations (Sarat-St. Peter, 2017), the ethical dimensions of extremist TTC—particularly regarding violence—provide an additional axis for consideration alongside the balancing of contradictory desires in challenging the technocratic state.

When TTC that incurs, deploys, or otherwise leverages violence is circulated through radical sharing, the challenges to the state can be at odds with—indeed a threat to—the politics of active equality, social justice, and human life. In her study of terrorist TTC, Sarat-St. Peter (2017) documents how the authors of homemade bomb-making instructions not only "equip unaffiliated readers who have no formal training in terrorist tactics with the capacity to attack" but also try to persuade them, through identification, to carry it out (p. 77). Failing that, though, the Jihadist instructions may generate enough fear in "alienated" readers, through the availability of materials and simplicity of the process, to mobilize support for the broader cause of "Muslims" via governmental policy or other institutional means (pp. 77, 87). Thus, in either case, from the instruction author's perspective, "Jihadist tactics always win—regardless of whether a reader follows the instructions" (p. 87). In the first case, the tactical narrative in the instructions, as Van Ittersum (2014) might describe it, aims to "present a character (the author) they may choose to identify with or become in some fashion, an experience they may want to have, a set of practices in which they may want to engage, and an object they may want to create and use in similar or different ways" (p. 236). Although Sarat-St. Peter (2017) does not elaborate on the second case, the alienation of the reader functions through a form of disidentification, a revulsive response to the person, the process, or the object of creation—and importantly the human destruction it entails. Put simply, through the viability of the tactical approach, the terrorist's instructions themselves generate fear, a type of terror performance. Such a response positions TTC beyond the instrumental: it is a tactic that gains credibility for being viable yet is ethically revulsive; it is functional in an extra-institutional sense but is abhorrent in its destructive effects. These unaffiliated, alienated readers play an especially important role in understanding extremist TTC, complicating the binary opposition of tactical communicators working against the strategic order of the state—particularly when coupled with the contradictory desires of fighting state violence with tactical violence.

Returning to the work of Colton, Holmes, and Walwema (2017), we can also consider extremist TTC in the violence directed against an oppressive state (e.g., fascist, totalitarian, plutocratic) that is itself violent. They highlight that "identifying a technical communication practice as tactical does automatically equate to moral justification" (p. 65) and that "tactics do not necessarily only affect those toward whom the tactic is directed" (p. 64). Their cogent example of the 2016 Malheur National Wildlife Refuge occupation in Burns, Oregon (a white militia's armed standoff against the US government) complicates the antistate/state binary by identifying other stakeholders, as the Malheur occupiers desecrated the land of the Northern Paiute Indians, disrupted operations of local public schools, and prevented

Bureau of Land Management workers—the area's largest employer—from doing their jobs (p. 64). On the one hand, the actions of the militia and its TTC documents (e.g., guidelines for kangaroo courts and "secret citizen panels") most likely alienated otherwise supportive audiences through "worry, hesitance, and even ire" (p. 65). On the other hand, the citizen engineering—the radical sharing emanating from Malheur—also resulted in circulating the spectacle as a prototype, a "dress rehearsal" for the violent insurrection at the Capitol in Washington DC on January 6, 2021 (see Bernstein, 2021; Rokala, 2021). Thus, as we consider militant TTC against the state, we need to account for both alienated and enthusiastic radical sharing audiences. When the insurgent violence reflects, indeed mirrors, the hegemonic state in terms of approach, personnel, or symbolic display, the analysis of TTC's ethical plane becomes invaluable not only for the case in question but also how it then circulates and what it generates.

The trajectory of violence may indeed determine a reception toward either alienation or radical sharing, and the degree or severity of the violence may depend on the militants' view toward the police order and its own violent systems. As Colton and colleagues (2017) have it, "[A]n order's degrees of repressiveness or oppressiveness can mean something entirely different depending on the perspective of the participants" (p. 62). With human lives at stake, a framework for ethics is particularly crucial, yet TTC scholars cannot look to de Certeau for productive guidance "to consider the ways in which we evaluate the ethics of a given set of tactical practices" (p. 59). Instead, Colton and colleagues (2017) offer a framework based on Cavarero's ethics of care useful "to characterize the totality of relations of those affected by a given tactical action, and, in turn, to attribute ethical behavior" (p. 60). As they explain, Cavarero's concept of human vulnerability is paramount, with the effects of TTC registered along a spectrum ranging from "care" to "wounding": "less a heuristic for determining exactly what are right and wrong actions and more a lens for questioning how one ethically justifies certain actions in relation to others" (p. 63). Furthermore, they clarify that, ultimately, an individual's *ontological singularity* is positioned on one end of the spectrum with "horrorism" on the other, with indices that range from care for the vulnerable individual at one end to ontological annihilation at the other: "[V]ulnerability functions as one way in which we can understand whether a given tactic cares for an individual's singularity and maintains vulnerability or, instead, works toward horrorism and the denial of singularity, the worst examples of the latter being extreme dehumanization through genocide or torture" (p. 65). I take up this framework of vulnerability, with its care-wounding spectrum, in my analysis of Weather, as a means of tracking TTC as it moves and repositions itself along that spectrum, showing us how an awareness of horrorism (and its threat to the individual and active equality) can influence TTC's trajectory.

Citizen Engineering: Days of Rage and the Weathermyth

Opposing the technocratic military and corporate presence on 1960s college campuses, SDS established a national reputation through its 1960s Port Huron Statement and propagation of New Left ideals—becoming the organization most closely identified with the Vietnam antiwar movement. Amidst the global protests of 1968 and taking center stage during the occupations of Columbia University, SDS leadership gave way to the militant Weatherman faction, which trained its members in street fighting tactics, organized itself into a network of semiautonomous cells, and eventually launched a sustained bombing campaign against government and corporate targets throughout the early 1970s.

The complex ideological debates within SDS/Weather are well documented by historians such as Varon (2004), Berger (2006), Eckstein (2016), and Rocha (2020), and while the organization's archival record is fragmented and incomplete, it reflects the typical organizational genres of correspondence, meeting minutes, event-planning documents, and an array of mimeographed flyers and lengthier position papers, many published through the newspaper *New Left Notes*. We could view early SDS through traditional lenses of technical communication in the context of its bureaucratic structure as a national political-lobbying network, but a different analysis is required to see how TTC emerges when SDS dissolves into an underground organization targeting the state. The TTC I consider here goes beyond Weather's original work with political tracts or protest proclamations (citizen journalism), focusing instead on the turn to a tactical use of violence and media spectacle (citizen engineering), with attention to the ethics of horrorism.

As SDS fell into disarray through internal conflicts—with one faction led by Weather leadership—it hoped to replicate the infamous violent police confrontations at the 1968 Democratic National Convention by returning to Chicago in October 1969 for the "Days of Rage." In the months leading up to the event, Weather openly advocated for the direct-action tactics of destroying property and provoking violence with the police, all documented in its campaign centered around a slogan to "Bring the War Home." One flyer, for example, exhorts protesters to action by "fighting against the pigs, the schools, and the courts," whereas another calls for "action not only against a single war or a 'foreign policy,' but against the whole imperialist system that made war a necessity" (see Students for a Democratic Society Records, 1958–1970). Such flyers—and other circulating documents—also decry capitalism, racism, schooling, and the war in Vietnam while advocating for freeing political prisoners and supporting the Black, brown, and other third-world liberation struggles at home and abroad, early examples of what Huiling Ding (2009) describes as a "rhetoric of proclamation," written by "anonymous rhetors who made short, dramatic, decontextualized, and often exaggerated proclamations to large, attentive, and receptive audiences" (p. 331). These prototypical, analog forms of radical sharing proliferated in the 1960s as weaponized media (see Lucas, 2019), and as historian Kirkpatrick Sale (1973) notes, SDS was a prolific contributor, thought of as "the 'writingest' organization around, and the prolificity is, considering the obstacles, amazing" (p. 125; emphasis in original). Furthermore, as McMillian (2011) explains, SDS provided a vital communication hub, and it "set the template for underground newspapers that functioned as open forums, to which virtually anyone could contribute. . . . Before long, underground newspapers in every region of the country began playing a similar role" (p. 14). With Weather in control of SDS, however, coordinated actions—over words—took priority as they called for action "based on the full participation and involvement of masses of people in the practice of making revolution; a movement with a full willingness to participate in the violent and illegal struggle" (Ashley et al., 1969, p. 9).

Leeman (1991) notes that Weather members were "the best known domestic terrorists" (p. 159), contending that the rhetoric of terrorist groups often mirrors, or reflects, that of the police order—or counterterrorist state. In his "bipolar exhortation," terrorist groups advocate the legitimacy of their cause as an equitable response to state violence, aiming to polarize "the body politic" by denying it a position of political neutrality and emphasizing "an inhumane system" (pp. 46–47). In this zero-sum game, action and violence is positioned as antithetical or superior to reasoned discourse or communication (pp. 47–56), so Weather—like any terrorist group—occupies a paradoxical stance, condemning the weakness of words to justify action but using words to do so: "Exhorting the body politic to act contains within itself the seeds of the contradiction" (p. 63). While Leeman (1991) rightly points out the contradictory desires inherent to elite vanguard discourse that proclaims populist revolutionary action (pp.

60–62), his larger description of Weather discourse suggests that terroristic rhetoric can largely be determined simply through (1) a group's opposition to the state and (2) a range of basic language features (metaphors, symbols, invective, etc.). Leeman (1991), however, ignores Weather's clearly articulated grievances against the state for its foreign-policy violence (influenced by private global corporations) and offers only a crude oppositional dynamic: "The Weathermen proposed to speak for, and about, the people. Directly opposing the people was the Imperialist-Capitalist-System. All three terms were used interchangeably and summarized a single entity: them" (p. 160). However, if Weather aimed to speak for anyone, it was "the people" of color and third-world populations, and if it aimed to speak to anyone, it was not an abstract "them" but specifically white America and its settler colonialism, all in the hopes of raising political consciousness and guiding a frustrated populace into tactical practice and agency beyond state-sponsored solutions—citizen engineering toward active equality.

Given Weather's stated intentions to provoke violence from and against the state, many moderates avoided the Days of Rage, resulting in only hundreds showing up in 1969, rather than the thousands from the year before. As Harold Jacobs (1970) notes, the October event gave Weather a national profile as extremist leadership: "The media created a Weather Myth: Weatherman soon became known as the most militant and omnipresent of white revolutionary organizations. . . . They projected a commitment on the part of white revolutionaries to move into total opposition to the state, regardless of personal cost" (p. 144). As the Weather vanguard led SDS into becoming the clandestine Weather Underground Organization, it staged one final public event to solidify the "weathermyth" and provoke the media into showing that Weather was indeed the threat it was perceived to be. The December event would either encourage the unaffiliated (through identification) to take up arms against the state or for the alienated (finding the tactics ethically revulsive or abhorrent) to support the broader cause of the Movement through policy advocacy or other institutional means.

Mirroring Wargasm, Weaponizing Whiteness

Weather promoted tactical practices over deliberation, instructing its foco-group cells to harden themselves for revolution. While most of the missives were verbal and designed to keep cells safely distant from the central organization, a key part of its ethos and militant formation was the negation of individual subjectivity in the name of the movement. As Jeremy Varon (2004) writes, in its efforts to strip itself of "bourgeois individualism," the cells were instructed to implement a set of community standards that rejected private property, required subsistence living, and denied basic leisure and creature comforts with all personal decisions sacrificed to leadership (p. 58). Two features of life in the collectives are worth noting here for their horrorist trajectory and threat to individual vulnerability. One took form through the "smash monogamy" campaign, which forced members to rotate sex partners, prohibited couples from exclusive relationships, and encouraged group sex to break down "counterrevolutionary" hang-ups (p. 58). The other took form through marathon Maoist "criticism/self-criticism" sessions, which "ostensibly sought to encourage political and emotional honesty and group bonding . . . to confront and root out their racist, individualist, and chauvinist tendencies" (p. 58). However, rather than communal enlightenment through care, individual vulnerabilities were exposed and exploited—indeed, with the aim to extinguish them—in sessions that were "part political trial, part hazing, part shock therapy, part exorcism, and, in a word used by more than one former member, part 'brainwashing'" (pp. 58–59). While not extending to the corporeal violence typified by horrorism, the ontological threat to individual subjectivity serves as one indicator of a horrorist ethic emerging in the name of Weather's ideological discipline and commitment.

While some supporters embraced the weathermyth, many observers thought that SDS had devolved into chaos and militancy, a role that Weather embraced as it coordinated a gathering in Flint, Michigan, in December 1969. Dubbed a meeting of "The War Council," it was a tactical event publicly promoted as a media spectacle: a "wargasm" and "an international youth culture freakshow." Flint served as a national call to arms and a rallying point for extremists: a spectacle designed to generate news headlines and direct attention before Weather went underground. As Sale (1973) writes, "It was without a doubt one of the most bizarre gatherings of the decade" (p. 626), which is quite an audacious claim for the psychedelic 1960s. A celebration of violence, the militant "happening" was a swan song for peaceful demonstration and public deliberation. As Varon writes, it was "a massive indulgence in symbols, a dizzying play of signs, mostly exhorting Weatherman's own members to more intense action. . . . [M]uch of the conference focused on the practical aspects of clandestine armed struggle, such as the choice of targets, the procurement of weapons, and the building of secure cells" (pp. 158–59). Amidst these TTC citizen engineering workshops, the lines between hyperbole and irony were difficult to define, with visual displays celebrating weaponry, sing-alongs about rioting, and the event's nadir when Weather leader Bernardine Dohrn praised the horrors of the recent Charles Manson cult murders (Dellinger, 1975, p. 152). With the My Lai massacre making headlines the month before (putting US military horrorism on full display), Weather staged its own horrific celebration of weapons, death, and human annihilation as "bringing the war home" transitioned from a protest-mobilization effort into a frenzied mission and vision statement for violence.

In materials distributed at Flint, Weather outlined a theory of armed struggle in "Everyone Talks about the Weather . . ." (1970), articulating its instructions to carry out the revolution from underground, empowered by the privilege and ubiquity of whiteness in the state:

> White revolutionaries live behind enemy lines. We are everywhere: above, below, in front, behind, and within. While we possess none of the machinery of the state, it is always close at hand. Our ability because we are white to move within the structure of the state, to locate ourselves in and around all of its institutions, opens up explosive possibilities for undermining its power. Our strategy must take into account the ways in which this particular asset can be used to prove material support for the strategy of the black colony. (p. 443)

It was preparing to work as a clandestine organization, the newly configured Weather Underground Organization, a dispersed network of militant cells to carry out internal revolution. In mirroring the ubiquity of the police order's whiteness, it would use tactics "above, below, in front, behind, and within . . . in and around" the state to undermine its power and to support active equality, particularly in Black America and universally for all oppressed peoples. Communications from SDS would no longer circulate from a central office but would emanate from tactical communiques tied to violent direct-action spectacles.

Explosive Revelations and Armed Propaganda

"Everyone Talks" (1970) specified a third course for the larger movement, one that avoided the growing apathy with peaceful protest yet did not devolve into the chaos of "nonstrategic terrorist activity" (p. 444). Instead, violent public spectacles against the state would inspire revolution:

"[M]ass demonstrations aren't going to bring down the state. They are one level of action and their continued usefulness lies mainly in their connection with a larger plan. The notion of public violence is increasingly key. That is, planning, organizing, and carrying off public and visible violent action against the state" (p. 444). Early 1970 witnessed bombing campaigns against judicial and police buildings across the country, many attributable to Weather collectives on both coasts and in the Midwest (see Burrough, 2015, pp. 92–102), but then an event that became known simply as "The Townhouse Explosion" escalated Weather's plans to go underground, forcing a reconciliation of their revolutionary TTC (and attendant horrorism) with their plans for militant action.

On March 6, 1970, three Weather members were killed while assembling a bomb inside a Greenwich Village townhouse in New York City, destroying the building and revealing an alarmingly large stockpile of explosives. Only fragments remained of the three dead bodies; two Weather members escaped the wreckage and went into hiding for the rest of the decade. The annihilation of bodies exemplifies Cavarero's (2011) account of horrorism being beyond terror: "The body undone (blown apart, torn to pieces) loses its individuality . . . dehumanizing and savaging the body as body, destroying its figural unity, sullying it" (p. 9). Faced with the horrific realities of their own deaths, Weather's ethical trajectory changed: away from a broadly imagined violent struggle that included human targets and toward a more clearly defined coordinated messaging campaign of "armed propaganda" that targeted only the property of the technocratic state.

Confronting the horrors of war truly at home, Weather members reimagined their tactics and decided on both a political and moral level, as Berger (2006) explains, "that armed propaganda—targeting only property—was a more appropriate level of struggle at the time" (p. 130). The aim was not just to adjust their own vision and actions but to discourage other militant groups from taking human lives: a change of course in its instructions for citizen engineering. Weather leader Bill Ayers (2001) recalls that "we dreaded the possibility of two, three, many Townhouses, and we hoped to use our celebrity in the lunatic left as well as the gathering Weathermyth in the larger world to persuade others to pull back" (p. 228). In a series of bombings that followed, Weather targeted government and corporate buildings only after ensuring that they were unoccupied or providing enough advance notice to evacuate personnel (see Burton-Rose, 2010, p. 36). Each incident of property destruction was tied to a communiqué that connected the act to the specifics of the revolutionary struggle, a combination characterized by some as "exploding press releases" (Burrough, 2015, p. 5) that demonstrated a militant genre for any individual to take agency against the state. As Weather operative David Gilbert (2012) recalls, "Every target had to be a clear symbol of the power structure that oppressed people. Every action was accompanied by a communiqué explaining it politically" (p. 162). Although its trajectory toward horrorism was fully on display in Flint, the horrors of the Townhouse recalibrated Weather's tactical ethics, revealing its own vulnerabilities within the organization and—ostensibly, at least—toward care for all humans, civilian or otherwise.

Inspired by the Tupamaros movement in Uruguay, Weather advocated for the tactics of armed propaganda: "transmitting political messages through violence of a spectacular and symbolic, yet measured, nature . . . to generate admiration and support for the perpetrator, rather than the fear inherent to terrorism" (Brum, 2014, p. 396). Armed propaganda actions were acts of symbolic violence on a grand scale, choreographed as tactical communicative events that challenged interpretive frameworks that might otherwise classify the crimes as vandalism, sabotage, or terrorism. Armed propaganda wasn't sabotage: it didn't expect to disrupt the state's military plans, unlike other movement groups that, for example, destroyed a draft-induction center to prevent young men from being sent off to war. The advance warnings, the explanatory communiqués, and consistent targeting of only state and corporate buildings were TTC choices that Weather legitimized as not-terrorism because the actions were not random, did not aim to hurt people, and did not create generalized fear among the public.

The first two bombings in June 1970 targeted New York Police buildings as "a symbol or institution of Amerikan injustice" ("Communique," 1970, pp. 510–11), followed by bombings at the Presidio army base in San Francisco and a Bank of America building in New York City. Armed propaganda was a form of TTC not only instructing other extremists but also sending messages about active equality and social justice: against the institutions that were harming lives on a regular basis, individually and on a massive scale. The first communiqué in July 1970, "A Declaration of War," contrasted the ineffectual efforts of mass demonstrations to the state violence that resulted in human deaths, both individual and en masse:

> Tens of thousands have learned that protest and marches don't do it. . . . The hundreds and thousands of young people who demonstrated in the Sixties against the war and for civil rights

> grew to hundreds of thousands in the past few weeks actively fighting Nixon's invasion of Cambodia and the attempted genocide against black people. The insanity of Amerikan "justice" has added to its list of atrocities six blacks killed in Augusta, two in Jackson and four white Kent State students, making thousands more into revolutionaries. ("Communiqué," 1970, p. 509)

This initial communiqué announced the New York bombing that followed a few weeks later to celebrate Black power figureheads "and all black revolutionaries who first inspired us by their fight behind enemy lines for the liberation of their people" ("Communique, 1970, pp. 510–11). Put simply, Weather's tactical destruction of property was used as a spectacular call for active equality: a coordinated messaging campaign to call attention to the state's destruction of human lives, particularly against people of color.

It has long been a commonplace that members of any civil, democratic society should not "condone" violence, but what is permissible in the name of dissent is not always clearly defined. Physical conflict with police resulting in personal injury is not the same as property destruction, though both are often cited interchangeably as examples of Weather "violence." The term *violence* serves as a convenient shorthand for contrast to what we might call the preferred tactics of nonviolence—marches, rallies, sit-ins—but our analyses of militant TTC should be informed by finer distinctions and grounded in deeper historical contexts. For example, in their analysis of the 1999 World Trade Organization protests in Seattle, DeLuca and Peeples (2002) define "symbolic violence" as "acts directed toward property, not people, and designed to attract media attention" (p. 256). And as philosopher Stephen D'Arcy (2014) concludes, material inconveniences to the state caused by such disruptions are similar: "Damaging property—at least replaceable property, and especially corporate property—usually only costs others money. . . . If that is the injurious part of the action, then it is neither more nor less injurious than costing people or businesses similar amounts of money by means of boycotts, picket lines, or other kinds of disruption" (pp. 116–17). As we consider the case of Weather and its broad range of TTC in the name of active equality, social justice, and outright revolution, we may not want to entertain the legitimacy of armed propaganda as a suitable option, but we must consider the ethical trajectories behind such tactics and how the escalation toward extremist acts places communicators dangerously close to the horrorist agendas that they hope to extinguish.

References

Ashley, K., Ayers, B., Dohrn, B., Jacobs, J., Jones, J., Long, G., Machtinger, H., Mellen, J., Robbins, T., Rudd, M., & Tappis, S. (1969, June 18). *You don't need a weatherman to know which way the wind blows.* New Left Notes.

Ayers, W. (2001). *Fugitive days: A memoir.* Beacon.

Berger, D. (2006). *Outlaws of America: The Weather Underground and the politics of solidarity.* AK.

Bernstein, M. (2021, January 8). *Armed occupation of Malheur refuge was "dress rehearsal" for violent takeover of nation's Capitol, extremist watchdogs say.* Oregonian. https://www.oregonlive.com/crime/2021/01/armed-occupation-of-malheur-refuge-was-dress-rehearsal-for-violent-takeover-of-nations-capitol-extremist-watchdogs-say.html

Brum, P. (2014). Revisiting urban guerrillas: Armed propaganda and the insurgency of Uruguay's MLN-Tupamaros, 1969–70. *Studies in Conflict & Terrorism 37*(5), 387–404. https://doi.org/10.1080/1057610X.2014.893403

Burrough, B. (2015). *Days of rage: America's radical underground, the FBI, and the forgotten age of revolutionary violence.* Penguin.

Burton-Rose, D. (2010). *Guerrilla USA: The George Jackson Brigade and the anti-capitalist underground of the 1970s.* University of California Press.

Cavarero, A. (2011). *Horrorism: Naming contemporary violence* (W. McCuaig, Trans.). Columbia University Press.

Colton, J. S., & Holmes, S. (2018). A social justice theory of active equality for technical communication. *Journal of Technical Writing and Communication, 48*(1), 4–30. https://doi.org/10.1177/0047281616647803

Colton, J. S., Holmes, S., & Walwema, J. (2017). From NoobGuides to #OpKKK: Ethics of Anonymous' tactical technical communication. *Technical Communication Quarterly, 26*(1), 59–75. https://doi.org/10.1080/10572252.2016.1257743

Communiqué #1 from the Weatherman Underground. (1970, 31 July). The Berkeley Tribe. In H. Jacobs, H. (Ed.) (1970), *Weatherman* (pp. 509–511), Ramparts.

D'Arcy, S. (2014). *Languages of the unheard: why militant protest is good for democracy.* Zed Books.

de Certeau, M. (1984). *The practice of everyday life* (S. Rendall, Trans.). University of California Press.

de Certeau, M. (1997). *The capture of speech and other political writings* (T. Conley, Trans.). University of Minnesota Press.

Dellinger, D. (1975). *More power than we know: The people's movement toward democracy.* Anchor Press/Doubleday.

Deluca, K. M., & Peeples, J. (2002). From public sphere to public screen: Democracy, activism, and the "violence" of Seattle. *Critical Studies in Media Communication 19*(2), 125–151. https://doi.org/10.1080/07393180216559

Eckstein, A. M. (2016). *Bad moon rising: How the Weather Underground beat the FBI and lost the revolution.* Yale University Press.

Everyone talks about the weather . . . (1969). In H. Jacobs (Ed.) (1970), *Weatherman* (pp. 440–47). Ramparts.

Gilbert, D. (2012). *Love and struggle: My life in SDS, the Weather Underground, and beyond.* PM.

Jacobs, H. S. (1970). *Weatherman.* Ramparts.

Kimball, M. A. (2006). Cars, culture, and tactical technical communication. *Technical Communication Quarterly, 15*(1), 67–86. https://doi.org/10.1207/s15427625tcq1501_6

Kimball M. A. (2017). Tactical technical communication. *Technical Communication Quarterly, 26*(1), 1–7. https://doi.org/10.1080/10572252.2017.1259428

Leeman, R. W. (1991). *The rhetoric of terrorism and counterterrorism.* Greenwood.

Lucas, B. (2019). Insurgent circulation, weaponized media: Waging the late sixties war within. In J. Ridolfo & W. Hart-Davidson (Eds.), *Rhet ops: Rhetoric and information warfare* (pp. 90–104). University of Pittsburgh Press.

McMillian, J. (2011). *Smoking typewriters: The sixties underground press and the rise of alternative media in America.* Oxford University Press.

Rocha, M. (2020). *The Weatherwomen: Militant feminists of the Weather Underground.* McFarland.

Rokala, J. (2021, January 7). *Statement: 2016 Malheur Wildlife Refuge occupation was a "dress rehearsal" for Capitol insurrection.* Center for Western Priorities. https://westernpriorities.org/2021/01/07/statement-2016-malheur-wildlife-refuge-occupation-was-a-dress-rehearsal-for-capitol-insurrection/

Sale, K. (1973). *SDS.* Random House.

Sarat-St. Peter, H. A. (2017). "Make a bomb in the kitchen of your mom": Jihadist tactical technical communication and the everyday practice of cooking. *Technical Communication Quarterly, 26*(1), 76–91. https://doi.org/10.1080/10572252.2016.1275862

Students for a Democratic Society Records, 1958–1970. Wisconsin Historical Society Library and Archives, Madison, WI.

Van Ittersum, D. (2014). Craft and narrative in DIY instructions. *Technical Communication Quarterly, 23*(3), 227–246. https://doi.org/10.1080/10572252.2013.798466

Varon, J. (2004). *Bringing the war home: The Weather Underground, the Red Army Faction, and revolutionary violence in the Sixties and Seventies.* University of California Press.

Chapter Fourteen

Hospitality at the End of the World

An Ideological Rhetorical Criticism of Tactical Technical Communication in *The Prepper Journal*

Ryan Cheek

What is hospitality at the end of the world? Derrida and colleagues (1984) write that "nuclear war is not only fabulous because one can only talk about it, but because [. . .] like all techné—the extraordinary sophistication of these technologies coexists, cooperates in an essential way with sophistry" (p. 24). A potent example of such cooperation between techné and sophistry is the online doomsday prepper community where technical advice about surviving the end of the world as we know it (TEOTWAWKI) when shit hits the fan (SHTF) is blended with extreme right-wing political beliefs and apocalyptic rhetoric. Scholars analyzing a range of prepper artifacts have noted connections between religious apocalypticism and political conservatism (Crowley, 2006), survivalists and white supremacy (Crockford, 2018), and doomsday preppers and modern consumerism (Foster, 2016). Using an ideological rhetorical methodology (Foss, 2018; Wander, 1984), I analyze the tactical technical communication (Kimball, 2006; 2017)—or TTC—found in *The Prepper Journal* (TPJ) to argue that doomsday preppers use TTC to construct an exclusionary postapocalyptic community where hospitality must be earned through the acquisition of survival resources and technical knowledge.

Pat Henry, a pseudonymic reference to the Revolutionary-era antifederalist Patrick Henry, founded TPJ in January 2013, hoping to provide a more realistic understanding of the prepper community and how to prep than the founder believed Nat Geo's hit show *Doomsday Preppers* did. Despite its name, it is more of a blog than a journal—its editorship is proprietary, the site is supported by ad revenue and a prepper store, and there is no defined level of peer review for posted articles. Readership can also quickly respond through comments on any given post. When the research was undertaken, there were about 1,795 articles posted with 12 million page views since the site was created. Most authors and commenters post pseudonyms to maintain OPSEC (operational security). Using acronyms in prepper culture can be viewed as an element borrowed from its paramilitary roots and conveys a sense of urgency that necessitates short and efficient communication practices.

This chapter analyzes the connection between TTC, apocalyptic rhetoric, and hospitality ethics. More specifically, ideological rhetorical criticism is employed to answer the following research questions: (1) Who does apocalyptic hospitality include, and whom does it exclude from the postapocalyptic imaginary? (2) What are the ideologies circulating in TTC-focused doomsday prepper communities? (3) How does TTC contribute to apocalypticism? Alongside significant technical instruction, articles in TPJ contain deeply conservative political rhetoric that undergirds apocalyptic hospitality defined by paranoia, social Darwinism, and radical individualism. Hospitality is not absent from this community; it is morphed by eschatologies of endism and electism deriding liberals and anyone outside the cis-hetero-norm as being too weak to survive the apocalypse while demonizing people of color and non-Christians as both causes of the end and continuing threats post-cataclysm. Spliced into the technical instruction of survivalist tactics are ideologies that convey apocalyptic hospitality that welcomes white, male, cis-heteronormative, Christian folks at the exclusion of difference. Like Jihadist TTC (Sarat-St. Peter, 2017), prepper TTC persuades through identification (Burke, 1969). The cognitive resonance of shared ideological commitments in a postapocalyptic world is a powerful medium for such identification. In the next section, I provide a brief multidisciplinary overview of research on apocalypticism to contextualize the present work.

Apocalyptic Rhetoric and the Ideology of Prepperism

Apocalypticism is an ancient and inevitable human phenomenon (Grove, 2015), but doomsday prepper culture has experienced a dramatic increase in the last decade, producing a uniquely American subculture (Mills, 2019). The scale and growth of the prepping industry in the United States have far surpassed similar phenomena worldwide, indicating a cultural dimension to prepping attributable to American society. Criminologist Michael Mills (2019) has argued that prepping is "fundamentally sustained by a precautionary logic" (p. 1272) responding to apocalyptic media scare-mongering saturating American culture. While prepper anxieties are "not premised on concerns exclusive to extreme right-wing ideology" (p. 1274), Mills's fieldwork exploring the rapid growth of prepping activities after the 2008 US election has "indicated that prepping is an overwhelmingly right-wing phenomenon" (p. 342). Similar and complementary perspectives may be found in many other disciplines.

Ideologies of white supremacy, vigilantism, and survivalism are linked by a general paranoia about the future, a distrust of the government, and a belief in radical individualism (Crockford, 2018). These ideologies are circulating more frequently because capitalism has picked up on their popularity and repackaged them for consumers in a phenomenon known as "Apocotainment" (Foster, 2016). Nat Geo's *Doomsday Preppers* television series is a prominent example of apocotainment that other scholars have analyzed and argued is characterized by apocalyptic masculinity (Kelly, 2016; Kelly, 2020) defined by paramilitarism, violent fantasies, and the rhetorical feminization of liberal collectivism. Hegemonic masculinity is leveraged as a survival tool for cis white men (Casey, 2016), living through a supposedly denigrated and emasculated modern existence.

Conservative politics is strongly motivated by religious apocalypticism (Crowley, 2006), but apocalyptic culture can be bipartisan. For example, concern over an impending Trumpocalypse motivated "liberal preppers" to seek refuge from an impending Trumpocalypse (Riederer, 2018). However, research has found that liberal and conservative preppers are motivated by different instigating scenarios. TPJ commenters have expressed their belief that liberals do not have the stomach or wherewithal to survive an apocalyptic event. Riederer (2018) notably points out that both groups miss the fact that marginalized folks already feel the impacts of apocalyptic events in their communities as they struggle with wealth inequality, severe environmental degradation, and the increasingly authoritarian xenophobic tendencies of supposedly liberal democracies around the world.

What is persuasive about doomsday prepping is that it offers an illusion of control over the chaotic nature of existence. The TTC found in TPJ is filled with this sort of imaginative game where an author posits a scenario and provides instructions on how to prep for that scenario. For example, an article might start by suggesting one risk in the apocalypse is the flood of folks who did not prepare to come to your door to rob, maim, and murder you and your family for your supplies. You can prepare for this scenario by following this checklist to ensure you have proper defenses against nonpreppers. The need for technical instructions is demonstrated by the existential risk that not being prepared in the scenario poses. Postapocalyptic imagination (Doyle, 2015) is easily identified in prepper culture, but there is another current underlying condition of community in play. The following section explains the performative entanglement between apocalypticism and hospitality ethics that produces prepper TTC in the TPJ and similar discursive communities.

The Apocalyptic Hospitality of Prepper TTC

Apocalyptic hospitality may be understood in at least two senses. First, apocalyptic hospitality is an openness to a bleak future that may never arrive. There is a skeptically revealed (Quinby, 1999) optimism in the TTC of preppers who believe they can out-prep the end of history. Second, apocalyptic hospitality is a closedness to the stranger at the end of time. An eschatology of electism pervades prepper discourse as the saved/worthy/prepped are juxtaposed against the damned/unworthy/unprepared. Hospitality in the apocalypse becomes instrumental and contractual as scarcity logic sets in. Ideology is more durable than physicality—anthropogenic ideologies are more likely to survive TEOTWAWKI than their embodied creators.

The performance of doomsday prepping, like the practice of hospitality, is an aporic negotiation between the unconditional unknowability of the apocalypse and the conditions of apocalypses that guide prepper theory and culture. Conditions of apocalypses and hospitalities are products of ideologies produced and reproduced through many textual iterations (e.g., novels, laws, movies, scriptures, games, manuals, etc.). The apocalypse, like nuclear war (an apocalyptic subset), is fabulously textual (Derrida et al., 1984)—that is to argue that the apocalyptic imaginary is sustained by text but also sublimely surpasses the limits of symbolic articulation.

Without the rule of law (WROL) is a common prepper term to describe an apocalyptic scenario where the security and comfort provided by systems of governance have been disrupted either temporarily (e.g., civil disorder) or permanently (e.g., revolution). It is also a conservative fantasy employed to scare the electorate into obedience. For example, bad-faith attacks on "defund-the-police" rhetoric often use enthymematic arguments to associate anarchic chaos with very reasonable critiques of societal overinvestment in policing nationwide. "Thin-blue-line" rhetoric is another referent to WROL, where police are framed as the boundary between order and chaos. This dynamic mirrors the aporia of hospitality that the philosopher Jacques Derrida once articulated as a metonym for ethics. Hosts require the potential for guests, without whom there can be no hospitality. Police require the potential for chaos, without which there is no law and order. Doomsday preppers require the potential for annihilation, without which they cannot survive the apocalypse.

Pointing out the tautology expressed in cultivating an ethic of hospitality, Derrida (2001) has written that "hospitality is culture itself and not simply one ethic amongst others" instead, "*ethics is hospitality*" (p. 16; italics original). For Derrida, ethics, as a theoretical-practical construct, is defined by negotiating an aporia between the unconditional, absolute law of hospitality and the conditional state laws of hospitality. Absolute or "pure hospitality consists in welcoming whoever arrives before imposing any conditions" (Derrida, 2005, p. 7)—those conditions will be imposed as the price of refuge and inclusion in the state apparat. Such conditions both constitute and fundamentally betray the principle of hospitality. As one international relations scholar puts it, "[F]or hospitality to be possible there must be an inviolable home, but that home must be constituted by closure *as the very possibility* of openness" (Bulley, 2006, p. 652). Hospitality is thus spatial and temporal since someone must first occupy a territory to welcome strangers into.

The spatial-temporal feature of hospitality also helps to define what hospitality is not: charity, conflict resolution, humanitarian aid, and so on. Bulley (2015) has argued that "hospitality does not take place in just *any* space; nor does it involve a transgression of *non-meaningful* boundaries" (Bulley, 2015, p. 189; italics original). The end of time(s) is a meaningful boundary prepperism ideally transgresses through acquiring knowledge and resources. The TTC of doomsday prepperism is an insufficient hedge against barely conceivable apocalyptic events resting at the edge of time. Nonetheless, doomsday prepperism is built on and propagated by communities sharing technical expertise premised on collective eschatological fantasies about TEOTWAWKI. Prepping attempts to reconcile the promise of revelation with the futility of survival are contradictory entailments of apocalyptic rhetoric. I use the word *attempt* because there is no possibility of reconciliation; it cannot occur just like absolute welcomeness toward the stranger may not be fully reconciled with the exclusionary boundaries that hospitality necessitates. Whether to welcome or prep, both actions imply mastery over space, place, and time. In the next section, I apply ideological rhetorical criticism of TPJ to answer the following research questions: (1) Whom does apocalyptic hospitality include, and whom does it exclude from the postapocalyptic imaginary? (2) What are the ideologies circulating in TTC-focused doomsday prepper communities? (3) How does TTC contribute to apocalypticism?

An Ideological Criticism of TPJ

Whom Does Apocalyptic Hospitality Include, and Whom Does It Exclude from the Postapocalyptic Imaginary?

Ideological rhetorical criticism (Foss, 2018) can help us to understand how the affective sublimity of fear helps to constitute doomsday prepper communities. Such analysis helps reveal the beliefs, values, and assumptions behind the community that constitutes TPJ because it is an influential blog that claims to be more representative of the prepper community than apocotainment artifacts like Nat Geo's Doomsday Preppers series. Rhetoric as methodological inquiry is vital to revealing ideology because, as McGee (1980) pointed out, "the falsity of an ideology is specifically rhetorical, for the illusion of truth and falsity with regard to normative commitments is the product of persuasion" (p. 4). Prepperism as an ideology is concealed through the technical commitments to preparation; that is, TPJ authors and commentators alike can frame prepping as an ideologically neutral act precisely because of its technical components, which make it more like a disciplinary endeavor than an ideological one. Prepper identification often occurs along political-ideological lines that may be expressed as shared exclusions—those for whom dominant doomsday prepper culture configures as responsible for their expendability and eventual demise. Technical language then becomes the basis for articulating such exclusions as implied audiences—who are not considered "prepper" and, therefore whose place in the postapocalyptic imaginary is unwelcome.

Philip Wander's (1984) concept of the third persona clarifies how the TPJ community defines its inside/outside boundaries. Where the first and second personas relate to the speaker and their audience, the third persona is the audience that is left out of the speaker's intention. They may be targeted and spoken about in a text; they are, however, not spoken to in that text. Morris III's (2002) fourth persona, or more specifically Gibbons's (2021) technologically updated persona 4.0, further theorizes the double audience—one who gets the coded message and another who does not. Search engines, for example, are algorithmic audiences that websites code for; websites operationalize a symbolic order largely unseen by their supposedly chief audience. In the third persona, the critic may find the excluded audience. In the fourth persona, the critic may find the duped audience. In analyzing the third and fourth personas of TPJ, I hope to reveal how ideologies at play in the blog help to construct who is and is not a prepper, which has significant implications for how preppers come to believe who does and does not deserve to survive the apocalypse.

Democrats and liberals are a third persona of prepperism conveyed through tactical technical discourses. There are many articles in TPJ dedicated to the political takes of preppers, and an incredible amount of them exhibit fear of the Obama administration, praise for the Trump administration, and a general disdain for liberals. For example, in an article about the need to retrain any surviving liberals after the apocalypse, John D (2016) has written: "Because you're reading this, you're probably not one of those delicate butterflies. You may ask, 'what does this pussification of young adults have to do with me?' It's simple; in a post-apocalyptic world, you'll have to live among them. They'll be the ones who'll come to you begging, when they realize they have no survival skills" (para. 6). In feminizing liberal culture, D affirms both the identity of preppers as conservative as well as their responsibility as properly masculine subjects to protect and re-train liberals to survive harsh post-apocalyptic realities. When TPJ users are not denigrating liberals as weak and feminine, they paint them as potential enemies that motivate their preparations. In response to one of the few articles welcoming liberal preppers to the site, a commenter named John Parker writes that "liberals are the REASON for prepping."

What Are the Ideologies Circulating in TTC-Focused Doomsday Prepper Communities?

In an explicitly political article written by Wild Bill urging readers to vote in the 2018 midterm elections, the pseudonymous one-time owner of TPJ wrote that "this election is about our survival [. . .] Please, put on your best prepper 'can do' attitude and walk, run, drive, crawl, hitchhike, paraglide, swim or skateboard to your polling place and do your civic duty" (2018b, para. 6). His framing of the election in such stark terms is not necessarily unique to preppers. However, the overt electoral call to action based on survivalism reveals a highly problematic ideology. Characterizing liberals, and more specifically Democrats, as threats to survival renders them political enemies and absolute enemies of humanity (Rasch, 2003) with all the violence that entails. The feminization of liberals as derisive rhetoric brings attention to another pernicious ideology at play in prepperism: apocalyptic masculinity.

Women are a third persona of prepperism. As noted earlier, the ideology of prepperism is coded by apocalyptic masculinity built on paramilitarism, a disposition toward violence, and feminization of others (Casey, 2016). Even the founder of TPJ acknowledges that most of his writers and readers are men. In one article dedicated to convincing prepper spouses to prep, Pat Henry makes explicit heteronormative and gendered assumptions that spouses are emotionally driven wives who place too much faith in hope. Henry (2013b) writes that although he has "tried to convince [his] spouse of the impending doom [. . .] she did not want to believe that anything is hopeless." A reader, Jon R, commented in solidarity with the article's sentiment, declaring that his "wife is reluctant to prep, [. . .] If SHTF tonight, she'd be a major liability" (para. 8).

Prepperism is not just for any man, though either; it is for the militarized and fit man who takes bodily conditioning and self-defense tactics seriously. Addressing the reader, TPJ contributor TekNik (2015) writes that when SHTF, "you're gonna have to get that lazy ass of yours from off the sofa" and that "a prepper should have similar physical capabilities as an active-duty soldier" (para. 2) An anti-fat ableist ideology (Mollow, 2017) is embedded in this quote and circulating in the TPJ literature more broadly. Demonstrating this visually, TekNik (2015) includes a meme of a large guy holding a gun with a top caption that reads "OBSESSED WITH PREPARING FOR APOCALYPSE" and a bottom caption that pithily finishes "CAN'T CLIMB STAIRS WITHOUT FAINTING" (all caps use from the source). The reality is that one's ability to survive an apocalyptic event has, more than anything, to do with luck than any level of preparedness or bodily fitness. Fat and disabled folks are another third persona in prepperism—get fit or die when SHTF.

Many in the TPJ community cherish the nuclear family ideal, with several articles dedicated to parenting pre- and post-apocalypse. In a popular article about prepper parenting, an anonymous (2018b) author argues that "kids are one of the main reasons why people turn to preparedness, and protecting and preserving a family is one of the main reasons why people are tuned into the idea of future-proofing their life" (para. 1). The heteronormative logic here is apparent: it is not enough to survive; you need to breed so that you may remake the world after its end—making queer life another third persona of prepperism. Perhaps even more disturbing is the brand of parenting that prepper ideology instills, as articulated in editor commentary that precedes a reader-submitted opinion piece about the type of disciplinary child-rearing prepperism demands; Wild Bill (2019a) writes that it is "a touchy subject perhaps in a world overrun with P.C. morality" but "the real world is basic to prepping and corporal punishment reflects the realities of the real world" (para. 1). The endorsement of parental violence indicates toxic fatherhood masculinity embedded within prepperism and its ideological defense of the nuclear family.

White-supremacist ideology is prominent in various TPJ scenario-planning articles about potential apocalyptic events. For example, TPJ contributor Keith Pounds (2014) advocates for racial profiling as a survival tactic, writing that "hospitals in the U.S. are under direct threat from Islamist extremist attacks" in part because of "foreign and domestic physicians who have become 'radicalized' by Islamist extremist ideologies" (para. 7). Another TPJ author using the name Capt. William E. Simpson (2014) bemoans the apocalyptic threat immigrants pose, writing: "In addition to illegal immigrants pouring into the U.S. over our southern border, a contingent of whom are known to be Jihadists are the people who are flying commercial airliners right into America from countries in. Africa where the Ebola virus is raging out of control" (para. 5). In years past, prior to the Obama open-border policy, these Jihadists would have had a far more difficult time gaining access to the American homeland. However, that is no longer the case.

By linking terrorism, immigrants, and disease as existential threats to the "American homeland"—a xenophobic phrase that parrots fascist rhetoric—Simpson demonstrates the racial coding of apocalyptic logic. It is also a prescient example given the massive COVID-19 pandemic that occurred under the self-styled champion of restrictive immigration policies of former President Trump. Although it is the disease that kills, rhetoric codes human beings as diseased vectors of transmission that pose an existential risk to the host populations—an association that makes uncomfortable partners of public health and anti-immigrant discourses circulating in prepperism and American culture more broadly.

In prepper ideology, faith is critical to human endurance and perseverance after the apocalypse. Preppers must attend to their (Christian) spiritual needs in hard times, as exemplified by the TPJ author Red J (2017), who, comparing the postapocalyptic with the Old Testament, wrote, "Life in a grid-down situation will become extremely difficult," so "some believers will feel like they've been abandoned or punished by God" (para. 1) The Bible enjoys prominence in TPJ as the only religious text mentioned and quoted from, acting as inspiration, motivation, and a rationale for doomsday prepping. Another article about homesteading before and after doomsday exemplifies this point, where Matt Sevald (2015) laments that he does not believe our current technological society is how we are meant to live; instead, he argues that "we're supposed to physically toil for our daily bread," and that desk jobs are "killing us as we spend 1/3 of each day doing it" as evidenced by deadly medical afflictions such as cancer, which Sevald believes is "a consequence for us deciding yet again to do things our own ('easier') way" (para. 6). In this frame, prepping is given a transcendent purpose, while the disease is a divine punishment for our technological dependence and separation from God's creation.

How Does TTC Contribute to Apocalypticism?

The fourth persona, or persona 4.0 (Gibbons, 2021), helps expose "the machine as covert rhetorical audience" (p. 52) or winks, which "arise out of the bind generated via the internet's attention economy, with its endless morass of material, the need for search engines to sort through it, and the resultant doubled audiences of human and machine" (p. 54). Table 14.1 attempts to reveal at least some persona 4.0 that animates TPJ by approaching the website as a search engine might, by examining the prevalence (or lack thereof) of search terms within the site to understand the ideological commitments of the TPJ community. Although keywords were subjectively chosen and are mere fragments (McGee, 1990) of a much larger textual cloth and rhetorical ecology (Edbauer, 2005), they are informative in characterizing a TPJ persona 4.0.

Articles featuring women often focus on adapting prepperism for women specifically, as in the case of Walken's (2021) headline above about bug-out bags for women. Although there are plenty of articles about choosing an appropriate bug-out bag, most do not mention identity characteristics such as gender as a factor. Prepper ideology does not assume women's participation; instead, prepperism must be translated for anyone who is not a cis-gender man. Other gender-focused TPJ articles are concerned with defining family roles in the postapocalyptic imaginary; for example, Henry (2014a) asks and answers:

Table 14.1. Search Keyword Frequency in TPJ

Keyword	#	Headline Example
Prepper	1721	Backups and Alternatives—A Preppers Mantra (Parris, 2019)
Men	1710	The 5 Wise Men of Prepping (M, 2016)
War	1332	How to Survive Civil War 2020 (Anonymous, 2018a)
Family	1055	8 Helpful Tips for Introducing the Prepper Lifestyle to Your Family (Anonymous, 2020)
Gun	617	Shotgun for Prepper: Be Ready for Home Defense, Riots, and Looters (Anonymous, 2018c)
America	494	Prepping—Is America Ready? (Johnson, 2020)
Weapon	471	The Weapon of Next-to-Last Resort (Bill, 2018a)
Military	351	A Second Civil War Is Not What You Have to Worry About (Henry, 2020b)
Tactical	255	The Tactical Mindset! (Wilson, 2016)
Killing	147	Survival Tools You Can Make out of Scrap (Tara, 2018)
Women	142	Best Bug Out Backpacks for Women (Walken, 2021)
Terrorist	126	Is the U.S. Government Building the Terrorist They Need? (Henry, 2015)
Violence	126	When the Music Stops—How America's Cities May Explode in Violence (Bracken, 2013)
Apocalypse	96	Apocalypse Training 101: Learn How to Survive the End of the World (Henry, 2016)
Fitness	81	Prepper Conditioning: Total Body Fitness for When SHTF (Thift, 2016)
Revolution	64	Are We Spinning out of Control? (Bill, 2019b)
Obama	42	Open Border, Ebola & ISIS: A Perfect Storm for America? (Simpson, 2014)
Bible	36	The Bible, Preparedness, and the 2nd Amendment (Anonymous, 2016)
Liberal	32	The Rise of Liberal Preppers—Welcome to the Party! (Henry, 2017)
Trump	29	Are You Ready to Be Called a Domestic Terrorist? (Henry, 2021)
Democrat	26	Democrats, Republicans, and NRA Work Together on Gun Confiscation (Henry, 2013a)
Elections	22	Do the Results of the Election Have You Prepping for Doomsday? (Henry, 2020a)
Conservative	20	Cupcakes and Conservatives (D, 2016)
Gender	18	Putting Women in Their Place When the SHTF (Henry, 2014a)
Republican	17	Now That the Dark Ages Are Returning (Bill, 2018c)

Source: Created by the author.

> What is a woman's place? I don't write or think about them in those roles though because like I said, I don't really ever want to see that future for them. [. . .] It doesn't mean I don't believe they are capable, but I know that wouldn't be their preference. [. . .] This has nothing to do with Political Correctness or gender equality and I am not trying to change traditional roles of men and women. I have a very conservative view of marriage and family and most things in life. (para. 9)

The passage checks many of the keyword terms used in table 14.1—conservative, gender, men, women, and family—that reveal an ideological commitment to the nuclear family (Cloud, 1998) in the TPJ community. Bolstering this claim is the fact that *family* is contained in about 59% of all TPJ articles. Although that does not necessarily mean all those articles are exclusively focused on the family, the construct is important in prepper culture.

TPJ's persona 4.0 is preoccupied with violence, weapons, and warfare—painting a decidedly bleak postapocalyptic portrait of life after catastrophic events. Much of the tactical advice found on the pages that return these search terms are given under the assumption of a return to a Hobbesian existence where brutality becomes a condition of survival. Pondering what constitutes a tactical mindset, TPJ author Wilson (2016) writes that "political correctness does not enter into it; we are talking about your life and death not banning super-size sodas or gay marriage." Wilson's existential framing reveals an ethic of expediency (Katz, 1992)—it is life and death, so we do not have time to debate tactical engagement. The pithy cultural asides also reveal a conservative ideological framing embedded in the author's conception of a tactical mindset. How might the stranger be met in such a violent postapocalyptic imaginary?

Prepperism demonstrates that hostility is as much a hospitality as any greeting at the threshold of the entrance (held space, place, and time). Greeting with a gun or a handshake is still a greeting (imprisonment is still hosting). Phenomenologically, the perversion of a thing must still be that thing; otherwise, the perversion would be unintelligible. Apocalyptic hospitality in this context is then figured as a preparedness to meet the stranger with violence because when SHTF, as TPJ contributor Tara writes, "[l]ooting is everywhere, and so are the killings. Law and order are noticeably absent. People can do whatever they please and going out is akin to suicide." The rest of Tara's article constitutes a prime example of prepper TTC, specifically on *bricolage* (Kimball, 2006). Common household materials may easily be turned into weaponry, like combining dried chili pepper and water to make a blinding pepper spray, turning a paracord bracelet into a bullwhip, and a flamethrower built out of a fire extinguisher. The irony of turning a safety device into a weapon that causes the specific type of devastation (fire) the original device was built to snuff out is a telling perversion; in the postapocalyptic imaginary, to be safe is to contain the potential for violence.

Finally, there is a politics to prepperism. That is, the TPJ community is not apolitical. Elections matter significantly to some of the most prominent contributors—including the founder of TPJ, Pat Henry. Part of the blog's work appears to be preppers' political activation. From broad criticisms of Obama that parrot conservative rhetoric about partisan issues such as guns and immigration to the gleeful and inauthentic welcoming of liberal preppers in the wake of Trump's election to the US presidency. Beyond the 2020 election, the TPJ founder's most recent work appears to be attempting to help preppers make sense of a post-Trump and post-COVID world. COVID was indeed apocalyptic in many senses of the term; for example, the devastation it wrought has also revealed systemic weaknesses tearing at the fabric of American society, such as massive income inequality and a severe lack of accessible quality healthcare nationally. For preppers, it was apocalyptic in these senses but likely in another as well; it revealed that apocalypses could be slow rather than fast and can generatively rebuild life rather than just entropically consuming it.

Conclusion

Multiple intersecting third and fourth personas exist at work in the TTC of TPJ. Hospitality toward the third persona is, at best neglectful but may easily be characterized as hostile. Prepperism as an ideology tends to configure anyone who is not a cis gender, white, fit, conservative male as a nonsurvivor that must be convinced, converted, or cast out from the postapocalyptic imaginary. Hospitality at the end of the world, like absolute or pure hospitality, may never be known because it requires a condition beyond symbolic imagination. However, that does not stop mediated representations of apocalyptic hospitality that indicate and contribute to many anthropogenic and ongoing apocalypses in the status quo. Apocalyptic hospitality is practiced in various forms today—the perversion of the welcome, the holding of the stranger at gunpoint, the hostage taking, and resource plundering of nationalistic survivalism operating on a global scale.

Prepperism may seem niche, but the eschatologies of endism and electism promoted in the TPJ reveal broader right-wing ideological trends that posit white Christian cis-men as victims of a changing world. Ethics is conveyed in prepper TTC, which often emphasizes the mutual exclusivity of preserving your own life and a stranger's life. Others in prepperism are either threats to be eliminated or burdens to shed. In synthesizing hospitality ethics with an ideological rhetorical analysis of prepper TTC, this chapter meaningfully contributes to several disciplines, such as technical communication and political rhetoric. Prepperism tends to enmity, and so then, too, do the objects of prepper instruction. Once institutions are decimated, no strategies are found, only tactical survival, where every decision could be your last.

There is a deadly gravitas in prepper TTC like that of jihadist TTC (Sarat-St. Peter, 2017). That may seem rather dramatic, but ideological fantasies about postapocalyptic survival are dramatic; living for the end is dramatic. However, it is also revealing in the most apocalyptic of senses. Stripped of societal mores and norms against murder and mayhem, the belief among mostly male-identifying preppers in the TPJ community appears to be that violent chaos will likely ensue in the apocalypse, which is a belief, whether correct or not, that is indicative of an inhospitable outlook that condones the hostile greeting. What happens when the apocalypse disappoints? When people who shouldn't survive do and when institutions thought to be destroyed endure? Post-COVID and post-Capitol insurrection, members of the TPJ are only now beginning to struggle with those questions—making it a fruitful area for future inquiry.

References

Anonymous. (2016, November 18). *The Bible, preparedness, and the 2nd amendment.* https://theprepperjournal.com/2016/11/18/bible-preparedness-2nd-amendment/

Anonymous. (2018a, March 17). *How to survive civil war 2020.* https://theprepperjournal.com/2018/03/17/survive-civil-war-america/

Anonymous. (2018b, May 31). *Prepper parenting—involving the kids.* https://theprepperjournal.com/2018/05/31/prepper-parenting-involving-kids/

Anonymous. (2018c, November 14). *Shotgun for Prepper: Be ready for home defense, riots, and looters.* https://theprepperjournal.com/2018/11/14/shotgun-for-prepper-be-ready-for-home-defense-riots-and-looters/193

Anonymous. (2020, April 23). *8 helpful tips for introducing the prepper lifestyle to your family.* https://theprepperjournal.com/2020/04/23/introducing-prepper-lifestyle/

Bill, W. (2018a, June 8). *The weapon of next-to-last resort.* https://theprepperjournal.com/2018/06/08/weapon-next-last-resort/

Bill, W. (2018b, November 6). *You voted right?* https://theprepperjournal.com/2018/11/06/you-voted-right/

Bill, W. (2018c, November 7). *Now that the dark ages are returning.* https://theprepperjournal.com/2018/11/07/now-that-the-dark-ages-are-returning/

Bill, W. (2019a, January 12). *Does corporal punishment better prepare one for the real world?* https://theprepperjournal.com/2019/01/12/does-corporal-punishment-better-prepare-one-for-the-real-world/

Bill, W. (2019b, June 12). *Are we spinning out of control?* https://theprepperjournal.com/2019/06/12/are-we-spinning-out-of-control/

Bracken, M. (2013, September 5). *When the music stops—how America's cities may explode in violence.* https://theprepperjournal.com/2013/09/05/music-stops-americas-cities-may-explode-violence/

Bulley, D. (2006). Negotiating ethics: Campbell, ontopology and hospitality. *Review of International Studies, 32*(4), 645–663. doi: 10.1017/S0260210506007200

Bulley, D. (2015). Ethics, power and space: International hospitality beyond Derrida. *Hospitality & Society, 5*(2–3), 185–201. doi: 10.1386/hosp.5.2-3.185_1

Burke, K. (1969). *A rhetoric of motives.* Berkeley, Los Angeles, and London: University of California Press.

Cloud, D. L. (1998). The rhetoric of <family values>: scapegoating, utopia, and the privatization of social responsibility. *Western Journal of Communication, 62*(4), 387–419. doi: 10.1080/10570319809374617

Crockford, S. (2018). Thank God for the greatest country on earth: White supremacy, vigilantes, and survivalists in the struggle to define the American nation. *Religion, State & Society, 46*(3), 224–242. doi: 10.1080/09637494.2018.1483995

Crowley, S. (2006). *Toward a Civil Discourse: Rhetoric and Fundamentalism.* Pittsburgh: University of Pittsburgh Press.

D, J. (2016, December 12). *Cupcakes and conservatives.* https://theprepperjournal.com/2016/12/12/cupcakes-and-conservatives/

Derrida, J. (2001). *On cosmopolitanism and forgiveness* (M. Dooley, & M. Hughes, Trans.). London and New York: Routledge.

Derrida, J. (2005). The Principle of Hospitality. *Parallax, 11*(1), 6–9. doi: 10.1080/1353464052000321056

Derrida, J., Porter, C., & Lewis, P. (1984). No apocalypse, not now (full speed ahead, seven missiles, seven missives). *Diacritics, 14*(2), 20–31. https://www.jstor.org/stable/464756

Doyle, B. (2015). The post-apocalyptic imagination. *Thesis Eleven, 131*(1), 99–113. doi: 10.1177/0725513615613460

Edbauer, J. (2005). Unframing models of public distribution: From rhetorical situation to rhetorical ecologies. *Rhetoric Society Quarterly, 35*(4), 5–24. https://www.jstor.org/stable/40232607

Foss, S. (2018). *Rhetorical criticism: Exploration and practice* (Fifth edition). Long Grove, IL: Waveland.

Foster, G. A. (2016). Consuming the apocalypse, marketing bunker materiality. *Quarterly Review of Film and Video, 33*(4), 285–302. doi: 10.1080/10509208.2016.1144017

Gibbons, M. G. (2021). Persona 4.0. *Quarterly Journal of Speech, 107*(1), 49–72. doi: 10.1080/00335630.2020.1863454

Grove, J. (2015). Of an apocalyptic tone recently adopted in everything: The anthropocene or peak humanity? *Theory & Event, 18*(3). muse.jhu.edu/article/586148

Henry, P. (2013a, March 13). *Democrats, Republicans and NRA work together on gun confiscation.* https://theprepperjournal.com/2013/03/13/democrats-republicans-and-nra-work-together-on-gun-confiscation/

Henry, P. (2013b, April 26). *How to convince someone about prepping.* https://theprepperjournal.com/2013/04/26/how-to-convince-someone-about-prepping/

Henry, P. (2014a, May 29). *Putting women in their place when the SHTF.* https://theprepperjournal.com/2014/05/29/putting-women-place-shtf/

Henry, P. (2014b, July 18). *The airlift of illegals: Importing poverty into the U.S.* https://theprepperjournal.com/2014/07/18/airlift-illegals-importing-poverty-us/

Henry, P. (2015, April 21). *Is the U.S. government building the terrorists they need?* https://theprepperjournal.com/2015/04/21/is-the-u-s-government-building-the-terrorists-they-need/

Henry, P. (2016, November 12). *Apocalypse training 101: Learn how to survive the end of the world.* https://theprepperjournal.com/2016/11/12/apocalypse-training-101-learn-survive-end-world/

Henry, P. (2017, May 30). *The rise of liberal preppers—welcome to the party!* https://theprepperjournal.com/2017/05/30/the-rise-of-liberal-preppers-welcome-to-the-party/ 207

Henry, P. (2020a, November 9). *Do the results of the election have you prepping for doomsday?* https://theprepperjournal.com/2020/11/09/election-prepping-for-doomsday/

Henry, P. (2020b, December 17). *A second civil war is not what you have to worry about.* https://theprepperjournal.com/2020/12/17/second-civil-war/

Henry, P. (2021, January 23). *Are you ready to be called a domestic terrorist?* Retrieved from https://theprepperjournal.com/2021/01/23/domestic-terrorist/

J, R. (2017, December 27). *How church life will change in a post-apocalyptic world.* https://theprepperjournal.com/2017/12/27/church-life-will-change-post-apocalyptic-world/

Johnson, R. R., & Ranney, F. (2002). Recovering techne. *Technical Communication Quarterly*, 11(2), 237–239. https://doi.org/10.1207/s15427625tcq1102_16

Katz, S. B. (1992). The ethic of expediency: Classical rhetoric, technology, and the holocaust. *College English, 54*(3), 255–275. https://www.jstor.org/stable/378062

Kelly, C. R. (2016). The man-pocalypse: Doomsday preppers and the rituals of apocalyptic manhood. *Text and Performance Quarterly, 36*(2–3), 95–114. doi: 10.1080/10462937.2016.1158415

Kelly, C. R. (2020). *Apocalypse man: The death drive and the rhetoric of white masculine victimhood.* Columbus: Ohio State University Press.

Kimball, M. A. (2006). Cars, culture, and tactical technical communication. *Technical Communication Quarterly, 15*(1), 67–86. doi: 10.1207/s15427625tcq1501_6

Kimball, M. A. (2017). Tactical technical communication. *Technical Communication Quarterly, 26*(1), 1–7. doi: 10.1080/10572252.2017.1259428

M, B. (2016, January 9). *The 5 wise men of prepping.* https://theprepperjournal.com/2016/01/09/wise-men-of-prepping/

McGee, M. C. (1980). The "ideograph": A link between rhetoric and ideology. *Quarterly Journal of Speech, 66*(1), 1–16.

McGee, M. C. (1990). Text, context, and the fragmentation of contemporary culture. *Western Journal of Speech Communication, 54*(3), 274–289. doi: 10.1080/10570319009374343

Mills, M. F. (2019). Preparing for the unknown . . . unknowns: 'doomsday' prepping and disaster risk anxiety in the United States. *Journal of Risk Research, 22*(10), 1267–1279. doi: 10.1080/13669877.2018.1466825

Mills, M. F. (2021). Obamageddon: fear, the far right, and the rise of "doomsday" prepping in Obama's America. *Journal of American Studies, 55*(2), 336–365. doi: 10.1017/S0021875819000501

Mollow, A. (2017). Unvictimizable: Toward a fat, Black, disability studies. *African American Review, 50*(2), 105–121. doi: 10.1353/afa.2017.0016

Morris III, C. E. (2002). Pink herring & the fourth persona: J. Edgar Hoover's sex crime panic. *Quarterly Journal of Speech, 88*(2), 228–244.

Parris, R. A. (2019, January 4). *Backups and alternatives—a preppers mantra.* https://theprepperjournal.com/2019/01/04/backups-and-alternatives-a-preppers-mantra/

Pounds, K. (2014, July 10). *Why terrorists will strike US hospitals.* https://theprepperjournal.com/2014/07/10/terrorists-will-strike-us-hospitals/

Quinby, L. (1999). *Millenial Seduction: A skeptic confronts apocalyptic culture.* Ithaca: Cornell University Press.

Rasch, W. (2003). Human rights as geopolitics: Carl Schmitt and the legal form of American supremacy. *Cultural Critique*, 120–147.https://www.jstor.org/stable/1354661

Riederer, R. (2018). Doomsday goes mainstream. *Dissent, 65*(2), 14–18. doi: 10.1353/dss.2018.0023

Sarat-St. Peter, H. A. (2017). "Make a bomb in the kitchen of your mom": Jihadist tactical technical communication and the everyday practice of cooking. *Technical Communication Quarterly, 26*(1), 76–91. doi: 10.1080/10572252.2016.1275862

Sevald, M. (2015, January 13). *The case for homesteading*. https://theprepperjournal.com/2015/01/13/case-homesteading/

Simpson, W. E. (2014, October 19). *Open border, ebola & ISIS: A perfect storm for America?* https://theprepperjournal.com/2014/10/19/open-border-ebola-isis-perfect-storm-america/

Tara. (2018, July 25). *Survival tools you can make out of scrap*. https://theprepperjournal.com/2018/07/25/survival-tools-you-can-make-out-of-scrap/

TekNik. (2015, December 30). *Fitness for Preppers*. https://theprepperjournal.com/2015/12/30/fitness-for-preppers/

Thift, S. (2016, August 23). *Prepper conditioning: Total body fitness for when the SHTF*. https://theprepperjournal.com/2016/08/23/prepper-conditioning-total-body-fitness-for-when-the-shtf/

Walken, K. L. (2021, January 20). *Best bug out backpacks for women*. https://theprepperjournal.com/2021/01/20/best-bug-out-backpacks-for-women/

Wander, P. (1984). The third persona: An ideological turn in rhetorical theory. *Central States Speech Journal, 35*(Winter), 197–216.

Wilson, O. (2016, September 26). *The tactical mindset!* https://theprepperjournal.com/2016/09/26/the-tactical-mindset/

Chapter Fifteen

Marx in the Digital Age

The Critical Role of Tactical Technical Communication in Contemporary Humanism

Sandy Brack

Technical communication, in its various iterations, is in all cases a tool of active epistemology. That is, technical communications are artifacts of knowledge work, validated through individual or collective action—the performance of a task. While these communicative artifacts can be epistemologically prescriptive or descriptive, they are in both cases primarily teleological: the result is more remarkable than the process. This privileging of product over process is particularly relevant in profit-driven consumer societies where the labor of production is viewed as incidental to the value of the consumer products (and profit) produced. This objective focus risks the exploitation of the human actants whose very identities are influenced by their expenditure of labor involved in production.

The formation of identity within consumer culture and its significance to tactical technical communication (TTC) begins with Marx's concept of estranged or alienated labor. Occurring across four progressive stages, alienated labor is itself a teleological process, the product of which is an objectified and commodified individual. While it is important to understand each stage of alienation in its own right, this chapter will focus primarily on the stage of alienation that relates to the relationship between creative labor and the formation of identity: the alienation from "species-being."

In his *1844 Economic and Philosophical Manuscripts*, Marx (1932/1988) asserts that a human being's most authentic expression of *self* (or species-being) occurs through the expenditure of one's own labor in the act of creative production: the making of *stuff*. Simply put, Marx argues that what we *do* is not separate from who we *are*. Products of creative labor are tangible representations of *self*: an internal abstraction manifested outwardly. An individual "duplicates himself not only, as in consciousness, intellectually, but also actively, in reality, and therefore he contemplates himself in a world that he has created" (p. 77). Our own creative production always ends with the satisfaction of some need, either practical or aesthetic, and requires the appropriation of resources in the external world. Free, conscious, creative activity is the ontological manifestation of our species-character, or human-ness. Individuals are the most *self*-fulfilled when they are intimately connected with their process (labor) and their product.

This relationship between labor, product, and identity is readily exemplified in the bevies of -core trends that gained significant momentum during and after the pandemic years of 2020 and 2021: cottagecore, gamercore, fairycore, craftcore, and many others. At the root of many of these -core movements are tactical do-it-yourself narratives that fundamentally support the expression of *self* through tutorials on sewing, home-remodeling projects, cooking, decorating, gardening, and many others. *Aesthetics Wiki*, an open-source cataloguing site, identifies and classifies over three hundred -core subculture aesthetics, each with a central emphasis on expression of identity. If what we *do* is not separate from who we are, then -core aesthetics are intimately bound to expressions of species-being: the labor of creation facilitating an expression of *self* and vice versa. Of course, -core aesthetics are not novel in their approach to and use of tactical modes of DIY communications. While several scholars to date have explored the genre, do-it-yourself technical communications are, I suspect, a largely untapped source of TTC study. DIY communications, such as those found in the vast arrays of -core communities, ensures that the labor of creation and the subsequent product are closely joined to the individual; that is, an individual is not alienated from their process or their product by controlling powers of production for monetary profit.

Alienation from process and product creates individuals who are alienated from their species-being (*self*); they become objectified and ultimately commodified. In a consumer-driven society wherein we produce and consume more products than ever before, alienation from labor and product, and especially from species-being is particularly relevant. Because of our rampant consumption, an incredibly high value is placed on material products, and a disparagingly low value is placed on the labor and laborer that produced them. Technical communication scholarship has long recognized the need to consider the relationship between controlling powers of production and the labor involved in producing commodities or other consumables. For instance, Miller (1989) reflected on the potential restrictiveness of organizational communication: "[T]he Marxist critique . . . raises questions that are particularly germane to technical writing, questions about whose interests a practice serves and how we decide whose interests should be served" (p. 65). When neither the product nor the labor involved in its production belongs to the laborer, it follows that the laborer becomes a commodity that has a negotiable value that is marginal compared to the profit generated by the laborer's products.

Regarding labor as incidental to the final product is particularly relevant to strategic forms of technical communication that uphold and proliferate the values of institutions of power. This relevancy was clearly demonstrated in Katz's (1992) analysis of organizational communication in the form of a memo that recommended "improvements" to the Nazi gas vans used prior to brick-and-mortar gas chambers and death camps. Katz analyzed how this artifact of organizational technical communication emphasized the expediency of product over the dehumanizing labor involved in its production and the human lives that the product was designed to eliminate. Katz's (1992) work demonstrated how organizational power is observed when one is alienated from species-being and absorbed into an institutional or ideological identity: "In any highly bureaucratic, technological, capitalistic society, it is often the human being who must adapt to the system which has been developed to perform a specific function, and which is thus always necessarily geared toward the continuance of its own efficient operation" (p. 270).

The negation of species-being by institutional or organizational communication has far-reaching implications as Katz (1992) argued that "technical writing, perhaps even more than other kinds of rhetorical discourse, always leads to *action*, and thus always impacts human life" (p. 259). While the teleological object of technical communications may represent an institutional, ideological, or cultural identity, it is the action or labor involved in creating those objects or executing tasks that is most significant to an expression of *self*; when both product and labor are controlled by hierarchies of power, an individual's authentic expression of *self* is compromised.

Having been so alienated from the modes of an authentic expression of self, how do we seek to deliberately actualize a sense of self or autonomous identity within consumer culture? The inevitable contemplation of oneself in an external world created by one's own hands is by alienation replaced by a dissociated *desire* to contemplate oneself in an external world. No longer is the contemplation of self a natural consequence of creative labor, but by the alienation of creative labor, it becomes something uncertain, something requiring a hermeneutical instrument by which to translate or define one's self. In other words, because we can no longer recognize ourselves in the products we make, we attempt to recognize ourselves in the products we buy. Consumer spending is an attempt to reconnect with, or express ourselves through, consumer commodities, even though they are not in every instance produced by our hands directly. Who we *are* is no longer expressed in what we *do*; instead, who we *are* is expressed by what we *own*.

Marx considered that when an individual is alienated from their creative work, social value in material objects is mistaken for inherent value: we forget that social elements determine value and instead believe that objects possess an intrinsic value on their own. Fetishism, argued Marx (1867/1990), is the recognition of social, human, or metaphysical qualities in an object as *if they were already there*, independent of human influence. Commodities can be viewed as autonomous in this way only when their inherent value as ontological manifestations of self is forsaken through alienation. We forget that people make things and acknowledge with a degree of objective distance merely that things have been made. When commodities are viewed as autonomous, their value can be negotiable, manipulatable, and exchangeable. Consequently, our inherent drive to recognize ourselves outwardly allows consumerism to view identity as exchangeable, negotiable, and manipulable; identity becomes manufacturable.

Convinced of an inherent, autonomous value in commodities, consumers choose to purchase within their economic means items that best satisfy their needs or wants: items that align most accurately with their perception of an inner *self*. Market research in consumer identity reveals that product branding is one of many communication strategies used to appeal to a sense of self and reinforce a manufactured identity. Chernev and colleagues (2011) claimed that "in addition to serving as an external signal, brands can serve to establish and confirm a consumer's self-concept and identity" (p. 67). Emphasizing the role of strategically communicated brand value in reinforcing consumer identity reveals the illusory nature of choice and is a persuasive technique that allows controlling powers of production regulation over which products consumers may choose, thereby controlling individual expression of identity.

A cycle of retroactive need (Horkheimer & Adorno, 1944/2002) is evident in the instability of manufactured identities. That is, the more we attempt to confirm our*selves* in the things we buy, the more unstable our manufactured identities become, and the more things we buy in an attempt to stabilize it. The industry responds in kind: "[T]he increased degree of product commoditization in the past two decades . . . has made brand associations—in particular, associations related to one's self-identity—an increasingly important source of brand value" (Chernev et al., 2011, p. 67). Control of consumer identity becomes a pressing drive of production on an economic scale. Consumers with commodified and manufactured identities are forever trying to buy back themselves and thus forever reinforcing the system that classified, organized, and unified their identities to begin with: "[T]he more strongly the culture industry entrenches itself, the more it can do as it chooses with the needs of consumers—producing, controlling, disciplining them" (Horkheimer & Adorno, 1944/2002, p. 115).

Challenging a Manufactured Identity: Consumption as Secondary Production

Our alienation from species-being demands that we attempt to contemplate ourselves in a world that we consume rather than in a world we create. However, rather than seeing ourselves in the things we buy, we see instead carefully tabulated and calculated economic and social power inscribed in our material goods and reflected back onto us. This inauthentic contemplation of self has long relied on the passivity of consumer identity—an identity curated by retroactive need—which can be reproduced many times over to validate and perpetuate the system that curated it. De Certeau (1984) recognized this passivity and suggested a more active consumer identity by redefining what it means to consume in the first place. Specifically, de Certeau's account sees techniques such as *bricolage* and *la perruque* as tactically active means of consumption, or secondary production. " 'consumption' . . . does not manifest itself through its own products, but rather through its ways of using the products imposed by a dominant economic order" (pp. xii–xiii). In other words, the act of consumption constitutes a form of production itself.

For de Certeau (1984) consumption becomes active rather than passive when creative means are employed to make use of products for ends other than originally intended to creatively satisfy practical needs or aesthetic desires. TTC scholars have begun to explore the acts of consumption through extra-organizational communications (Kimball, 2006, 2017; Van Ittersum, 2014; Sarat-St. Peter, 2017; Getto et al., 2014; Colton et al., 2016; Edenfield, Holmes, & Colton, 2019; Mueller, 2019; Edenfield & Ledbeter, 2019) and have found that secondary production privileges individual agency over the products of consumption. Active consumption, or "secondary production" privileges the expenditure of labor involved in a consumer's relationship with their products. This privileging of labor over product is a tactical response to a strategically organized and centrally unified consumer culture that prefers the prescriptive use of its products through vertical technical communication; secondary production enacted through TTC is, therefore, a direct challenge to the alienation of labor as defined by Marx. Secondary production shifts the focus away from identifying one*self* in the products we consume to the ways in which we consume them: "[T]he consumer cannot be identified or qualified by the . . . commercial products he assimilates: . . . there is a gap of varying proportions opened by the use that he makes of them" (Marx, 1988, p. 32). Our species-being, or authentic expression of self, begins to reemerge as a consequence of tactical, creative labor.

When labor is privileged—especially the tactical labor of use—the product reflects the creator's authentic expression of self instead of a dominant social and economic order. Further, the narrative surrounding the creation and/or use of the product validates not only the product itself but also the author who created it (Van Ittersum, 2014); narrative validation usurps retroactive need. To return for a moment to the tactical nature of -core communications, mainstream and social-media platforms have repeatedly referred to the cottagecore movement as reactionary to consumer-driven capitalist societies (Reggev, 2020; Slone, 2020; Tiffany, 2021; Wodzińska, 2021; Gold, 2020). They acknowledge that the aesthetic is less about the tangible products associated with the aesthetic and more about the use to which each individual puts them. While the reconciliation of species-being is a clear result of the tactical labor involved in secondary production, the concept itself provokes a lingering question: What happens to an authentic expression of identity when our consumer products break down and we can no longer use them?

Reuniting with Species-Being: Tertiary Production and Its Challenge to the Dominant Economic Order

The aforementioned scholarship in TTC has primarily (while implicitly, in some cases) addressed the expression of self through secondary production techniques—bricolage: using existing products to create something new; *la perruque*: using existing products for means other than originally intended—but how can a similar expression be achieved when our consumer products no longer work? Expanding on de Certeau's (1984) concept, the present work advances a theory of tertiary production wherein individuals maintain their authentic expressions of *self* by inserting their creative labor into products they already own by way of repair. If secondary production focuses on the labor of use as emancipatory from Horkheimer and Adorno's (1944/2002) culture industry and reconciliatory to Marx's (1988) alienation from species-being, tertiary production serves these functions by the act of repair. Where secondary production privileges the labor of use, tertiary production privileges the labor of repair. Departing from the secondary productive techniques of bricolage and *la perruque*, tertiary production seeks neither to create something new out of existing products nor to make use of the broken product for a means other than intended. Rather, tertiary productive techniques reappropriate bricolage and *la perruque* in the maintenance or restoration of the original intended use of a product. Here, creative labor involved in a product's repair *re*-creates it, with the user having inserted a space for themselves into the product without fundamentally altering its functional capabilities.

Our technological devices are not only highly complex but are also increasingly significant sources of our identities. These devices often contain a great deal of what constitutes our identities: our discourse with friends and family; our social and professional calendars; our photos; our home videos; our travel itineraries; our social-media accounts; our workout routines; our music; our access to local and world news; our media entertainment; our shopping lists; our access to online shopping; and on and on. Because so much of our identity has been encapsulated in our cellphones, the malfunctioning of one can be devastating. Organizational communication seeks to capitalize on this devastation by controlling how, when, and where the device may be repaired. This organizational control of access to repair is not limited to cellphones but encompasses an array of complex technological devices, including industrial and agricultural equipment. In response to this organizational restriction of access to repair information, the open wiki site iFixit.com promotes and facilitates tertiary production or the tactical repair of one's own devices. In addition, iFixit.com is active in the social and political fight against organizational restriction to technical documents of repair ("Policy Objectives," n.d.), indicating the pressing shift in the desire to reunite with one's species-being against legally sanctioned, restrictive technical documentation.

The restriction of repair information by manufacturers necessitated an explosion of tutorial videos and guides and saw the do-it-yourself repair movement gain unstoppable momentum; it may not be unreasonable to assume that a growing number of people within consumer culture will search the internet for these tactical solutions before they will attempt to find an organizational how-to manual. One of the most significant websites focused on tertiary production is iFixit.com. In addition to its social and political activism, this site provides thousands of user-generated guides that instruct the viewer, step-by-step, on the repair of countless consumer commodities and endeavors to create, with contributions from consumers themselves, free repair guides for any electronic device ("Self-Repair," n.d.). The culture industry, by manufacturing identities through unified production and retroactive need, has facilitated and even encouraged our problematic throw-away consumer culture. Because individuals are coerced into buying back their identities through consumer goods, rather than risking the expenditure of creative labor by their customers (and thus a decrease in profit gained from alienated labor), manufacturers restrict the availability of repair information, actively dissuading an individual from fixing their own goods in favor of seeking repair at manufacturer-authorized repair shops. In some cases, consumers may be more likely to discard a product and purchase a new one instead of going through the hassle of authorized repair. Not only does tertiary production, therefore, have objective benefits to areas like economics and sustainability, but individuals may be inclined to use self-repair websites and participate in the creation of tutorials because the creative labor involved in doing so aligns them with their species-being and facilitates an authentic expression of self. With tactical repair guides available to consumers on sites like iFixit.com, individuals have an alternative to buying their identity; if identity is realized through creative labor, tertiary production allows an individual to realize their species-being by re-creating their products through the tactical labor of repair.

The process of re-creation was recognized and thoroughly explored by Getto and colleagues (2014). As part of iFixit's Technical Writing Project, which partners with 85 universities world-wide to teach technical writing in a hands-on capacity ("Welcome to," n.d.), the role of tertiary production and its relationship to identity is demonstrated through the creation of procedural discourse in the form of tactical how-to guides produced by small groups of undergraduate students: "[W]e attempted to understand each actor in terms of how it interacted with other actors within the network of the Technical Writing Project on both an epistemological (knowledge-making) and ontological (material) level" (p. 191). Implicit to this study is the expression of species-being through the creative labor involved in repairing (*re*-creating) technological devices given to the students; that is, how they creatively engaged both the technological devices given to them for repair and the unfamiliar tools used to aid the repair. Of significance here is not only the students' epistemological process of composing their how-to guides but also the ontological process that both preceded and succeeded it by seeing an expression of *self* in their creative labor and subsequent products. At the project's completion, the students did not fundamentally alter the function of the device, but through the shared creation of procedural discourse and an unbridled exploration of the tools provided to complete the project, the students re-created the device at hand by inserting themselves, by way of their creative labor, into it.

Implications

Providing an avenue for authentic expressions of self as achieved through one's labor of use, collaborative TTC projects such as iFixit's Technical Writing Program are ideal pedagogical techniques in classrooms that continuously strive to improve equity and inclusivity. As reconciliatory to the alienation of species-being in a culture that is increasingly focused on hyperproduction and rampant consumption, TTC, by means of secondary and tertiary production, holds significant implications in Marx's fourth alienation: of individuals from each other. The loss of authentic expression of self leaves an unclear path toward a relation to others who are also alienated from their own species-being (Marx, 1932/1988). In Western societies, an unstable, manufactured, and inauthentic expression of self is the teleological result of strategic forms of technical communication that prescriptively or descriptively promote a culture industry that sees individuals as consumers—consumers who unwittingly participate in their own marginalization.

Socially, "marginality" typically considers issues of gender, class, and race; TTC sees the term reflecting these identity markers as further appropriated by a culture industry that views individuals primarily as consumers: "Marginality is today no longer limited to minority groups, but is rather massive and pervasive; this cultural activity of the non-producers of culture . . . remains the only one possible for all those who nevertheless buy and pay for the showy products through which a productivist economy articulates itself. Marginality is becoming universal" (de Certeau, 1984,

p. xvii). Consumerism and the culture industry that drives it transcends gender, race, class, and age; TTC theory considers how division occurs between producers and passive consumers.

Our separation from each other concerns the culture industry insofar as how the nuances of individuality, or spectrums of differentiation, can be economized. Differences may be emphasized through practices such as product branding in order to encourage the purchase of certain products by statistically grouped and targeted individuals. Consumerism encourages individuals to focus on their individuality so that they may purchase just the right thing to express themselves. However, when "individual differences" are tabulated, calculated, unified, and perpetuated as qualifiers of self-expression by top-down organizational communication, they too often become lumped into contrived stereotypes. This focus on difference encourages the formation of misleading assumptions made about "others." Collaborative classroom projects that focus on TTC and explore the producer/consumer cultural relationship can perhaps work to challenge stereotypes that often germinate by such superficial evaluations as a person's personal possessions.

Strategic forms of technical communication such as targeted ad campaigns may perpetuate social division as they seek to capitalize on attempts at self-expression. Indeed, consumers buy products to show their identity as part of (or in support of) any number of things: the LGBTQIA+ community; an ethnic heritage; a political affiliation; a familial tie. However an individual may identify, there is a bumper sticker, a T-shirt, a lanyard for it. The print-on-demand online marketplace Redbubble encourages individuals to "find your thing" ("Our Mission," n.d.) and purchase a product that exactly reflects an inner part of one*self*. More indirectly, people may express their class identity by buying consumer products that are designer or off-brand, that are newly upgraded, or by how likely they are to attempt to repair a damaged product versus purchasing a new one. There are styles and colors of clothing that, however arbitrarily, enunciate one's affiliation with mainstream cultures or with sub- or countercultures.

Because consumer culture encourages us to explore and express ourselves materially, we consequently understand that others express themselves through their material possessions as well. Collaborative TTC projects seek to resolve this alienation of individuals from each other by emphasizing that self-expression arises not from the consumer products themselves, but from the use to which they are put, and especially from the unique situations that necessitate such use.

In technical-writing classrooms, students who are encouraged to think tactically about the past, present, or future use of consumer products and about situations that may necessitate such use will explore a connection to their species-being as precipitated by the agency afforded by secondary or tertiary production. The exploration of and reunion with an authentic expression of self introduces a stable identity, a locus from which students can confidently and openly relate to and validate the self-expression and lived experiences of others and view these as valuable contributions to the tactical navigation of a hegemonic culture of producers and consumers that is maintained by strategic forms of communication.

Equity and inclusivity are heightened as the result of encouraging a reconciliation with species-being and a resolution to the alienation of individuals from each other. Teachers of TTC may enrich their inclusive pedagogy by considering collaborative projects that emphasize either secondary or tertiary production. As enunciated in the 2014 study by Getto and colleagues, a single collaborative group project may be completed over the course of the semester; alternatively, several smaller projects may be undertaken with a change in group-member assignments each time. In either case, the focus would always be on the tactical use or repair of consumer products and on the various applications of de Certeau's (1984) tactical bricolage and/or *la perruque*. Students may be encouraged to repair anything from consumer tech (as in iFixit) to nontechnological household items. As students would be required to document their steps, they would include an explicit recognition of each group member's unique contributions along the way. Students would be further required to submit this documentation for review by the instructor at intervals to ensure successful completion of learning outcome objectives and to periodically allow students to reflect on their interpretation of the significance of the project at hand, of their own contributions to the group, and of the contributions of their group members. At all stages, students should be reminded that the focus of secondary and tertiary production is not on what they own, which can often be a point of socioeconomic division, but how they use what they own, emphasizing the potential for students to connect with each other experientially, by sharing themselves as they share their creative labor.

References

Chernev, A., Hamilton, R., & Gal, D. (2011). Competing for consumer identity: Limits to self-expression and the perils of lifestyle branding. *Journal of Marketing, 75*(3), 66–82. https://doi.org/10.1509/jmkg.75.3.66

Colton, J. S., Holmes, S., & Walwema, J. (2016). From NoobGuides to #OpKKK: Ethics of anonymous' tactical technical communication. *Technical Communication Quarterly, 26*(1), 59–75. doi: 10.1080/10572252.2016.1257743.

de Certeau, M. (1984). *The practice of everyday life.* University of California Press.

Edenfield, A. C., Holmes, S., & Colton, J. (2019). Queering tactical technical communication: DIY HRT. *Technical Communication Quarterly, 28*(3), 177–191. doi: 10.1080/10572252.2019.1607906

Edenfield, A. C., & Ledbeter, L. (2019). Tactical technical communication in communities: Legitimizing community-created user-generated instructions. *Proceedings of the 37th ACM International Conference on the Design of Communication, 37,* 1–9. https://doi.org/10.1145/3328020.3353927

Getto, G., Franklin, N., & Ruszkiewicz, S. (2014). Networked rhetoric: *iFixit* and the social impact of knowledge work. *Technical Communication, 61*(3), 185–201.

Gold, T. (2020, May 4). *The wicked truth about cottage core.* UnHerd. https://unherd.com/2020/05/the-wicked-truth-about-cottage-core/

Horkheimer, M., & Adorno, T. W. (2002). The culture industry: Enlightenment as mass deception (Trans. E. Jephcott). In G. S. Noerr (Ed.), *Dialectic of Enlightenment: Philosophical Fragments* (pp. 94–136) Stanford University Press. (Original work published 1944).

Katz, Stephen B. (1992, March). The ethic of expediency: Classical rhetoric, technology, and the Holocaust. *College English, 54*(3), pp. 255–275. http://www.jstor.org/stable/378062

Kimball, M. (2006). Cars, culture, and tactical technical communication. *Technical Communication Quarterly, 15*(7), 67–86. https://doi.org/10.1207/s15427625tcq1501_6

Kimball, M. (2017). Tactical technical communication. *Technical Communication Quarterly, 26*(1) 1–7. https://doi.org/10.1080/10572252.2017.1259428

Marx, K. (1988). Estranged labor (M. Milligan, Trans.). In *Economic and Philosophic Manuscripts of 1844 and the Communist Manifesto* (pp. 69–84). Prometheus Books. (Original work published 1932)

Marx, K. (1990). The fetishism of the commodity and its secret (B. Fowkes, Trans.). In *Capital: A Critique of Political Economy: Volume 1* (pp. 163–177). Penguin Classics. (Original Work published 1867)

Miller, C. R. (1989). What's practical about technical writing? In B. E. Fearing & W. K. Sparrow (Eds.), *Technical writing: Theory and practice* (pp. 61–70). Modern Language Association of America.

Mueller, D. (2019). The transmedia workbench: Technical communication and user-driven innovation. *Technical Communication (Washington), 66*(3), 244–256.

Our Mission. (n.d.). *Redbubble.* https://shareholders.redbubble.com/site/about-us/our-story

Policy Objectives. (n.d.). *The Repair Association.* Retrieved April 26, 2021. www.repair.org/policy

Reggev, K. (2020, October 21). *What exactly is Cottagecore and how did it get so popular?* Architectural Digest. https://www.architecturaldigest.com/story/what-exactly-is-cottagecore

Sarat-St. Peter, H. A. (2017). "Make a bomb in the kitchen of your mom": Jihadist tactical technical communication and the everyday practice of cooking. *Technical Communication Quarterly, 26*(1), 76–91. doi: 10.1080/10572252.2016.1275862

Self repair. (n.d.). *iFixit.* www.ifixit.com/info/what

Slone, I. (2020, March 10). *Escape into Cottagecore, calming ethos for our febrile moment.* The New York Times. https://www.nytimes.com/2020/03/10/style/cottagecore.html

Tiffany, K. (2021, February 8). *Cottagecore was just the beginning.* The Atlantic. https://web.archive.org/web/20210427003538/https://www.theatlantic.com/technology/archive/2021/02/aesthetics-wiki-cottagecore-tumblr-tiktok/617923/

Van Ittersum, D. (2014). Craft and narrative in DIY instructions. *Technical Communication Quarterly, 23*(3), 227–246. doi: 10.1080/10572252.2013.798466

Welcome to the ifixit technical writing project. (n.d.). *iFixitEDU.* Retrieved April 26, 2021, www.edu.ifixit.com

Wodzińska, A. (2021, January 21). *Cottagecore as a budding anti-capitalist movement.* Institute of Network Cultures. https://networkcultures.org/blog/2021/01/21/cottagecore/

Chapter Sixteen

The Narrative Construction of Social Justice in Technical Communication Pedagogy

Tracy Bridgeford

In 2006, when Miles Kimball (2006) introduced the idea of tactical technical communication (TTC) into our conversations, he opened the door to opportunities for incorporating and questioning user-generated content in research and pedagogy. For my purposes, I welcome the opportunity to interrogate my pedagogy because it enables me to more explicitly engage students in social-justice issues. My pedagogy has always involved what I have called "narrative ways of knowing" (Bridgeford, 2004), a concept I derived from educational psychologist Jerome Bruner (1991) who says that "we organize our experience and our memory of human happening mainly in the form of narrative" (p. 4). Unlike "logical and scientific procedures," Bruner says that the narrative form "can only achieve verisimilitude" and "whose acceptability is governed by convention and 'narrative necessity' " (p. 4). In other words, we construct our own realities through narrative. I believe every document is and tells a story, and writing a technical document is no different. It is only the style that differs. We contextualize a "reality," provide evidence we invent, and begin and end with a message of some kind. From this perspective, I centralize narrative in my assignments as a context for engaging students in a story within which they will act as technical writers. In doing so, students see the character's thoughts and actions, as well as their own, in what I think of as a *dual empathy*, a situatedness that enables learning. Narrative is what enables us (humanity) to experience dual empathy, revealing a path to TTC, which is how I envision this pedagogical approach.

Like Kimball (2006), I draw on de Certeau's (1984) notion of tactics to recreate my pedagogy that situate students in ways that enable agentic action. De Certeau examines tactics first in relation to *strategies*, that is, the formal institutional structures that establish power within in a community, such as the laws that govern traffic, and on the other hand, emphasize the way individual users circumvent those power structures in everyday practice. De Certeau (1984) describes two kinds of tactics, *bricolage*, or "making do," and *la perruque*. Given the context I provide for assignments, I am most interested in *la perruque* because it is a means for "resistance against authoritarian control" (Kimball, 2006, p. 81) that operates "outside, between, and through" institutional structures, at times resisting or subverting them (Kimball, 2006, p. 67). One of the ways Kimball (2006) applies de Certeau's concepts is by characterizing them as communal and local technological narratives, which tell a different story of a technology's intended use. Communal narratives (strategies) tell the story of "ideal narratives of how machines, technologies, and processes *should* work (manuals, procedures), could work (proposals), or did work (reports)" (p. 73; italics mine), while local narratives tell stories about how they *actually* worked (tactics). Local narratives situate users, Kimball argues, as "producers of documents and artifacts that subtly resist authority" (p. 82). For example, in Julian Orr's (1996) study of Xerox repairpersons, he followed a group that met once a week for breakfast to share local narratives about their experiences repairing copiers that the manual didn't or couldn't address, which could easily be thought of as contributions to a "shared tactical story" (Kimball, 2006, p. 74), or user generated content, in resistance to the confines of the official printed manual. As my discussion of a narrative construction of social justice will show, I make use of both *strategies* and *tactics*, creating opposing forces that situate students within "a fictional space" that encourages tactical action (de Certeau, 1984, p. 79), albeit imaginatively.

The purpose of this chapter is to describe both the transformation of my own sensibilities first as they affect my pedagogical thinking through a lens of tactics, which involves the use of a narrative context (science fiction), through the lens of what Walton, Moore, and Jones (2019) call the three Ps—positionality, privilege, and power. Following that self-reflexive discussion, I consider conversations from the last decade about technical communication and the need for more explicit social-justice pedagogies in order to situate my transformation within this conversation. I then recount a short history about narrative and technical communication pedagogy from as early as the 1980s. The last part of the chapter focuses on the novel I use as a context, *Terrarium*, and the imaginary journey I ask students to take in order to develop agency. Using a tactical lens, I explore the assignment scenarios I used for assignments based on the narrative context. I have argued for more than 20 years that this notion of agency both creates and acts as part of the narrative context to make possible the technical content assigned to students. Given the novel I use as a context, I found it challenging at first to instill a social-justice perspective into the story given its focus on a totalitarian society, but, as I discover upon conclusion of this reflection, I wasn't doing what I thought I was doing.

I structure this chapter as part memoir (à la Angela Haas, 2012), part disciplinary conversation about pedagogy, narrative, social justice, and TTC to capture the moments of insight I experienced throughout last year. Originally, my concept of technical communication pedagogy grew out of an understanding of Carolyn Miller's (1979) humanistic rationale as decidedly focused on how humans work together through (although she doesn't use this word) narrative, or talking to each other as part of a team as they go about the practice of a community. This humanistic foundation has enabled further and further introspection in the field, opening up spaces for ensuing conversations about ethics, social construction, cultural studies, diversity, and social justice—all of which tell stories of how humanity interacts in practice. Students come into our classrooms from many fields, many of which rely on the logical and scientific explorations of knowledge. But it is narrative that enables them to share and interpret that knowledge. Miller (1979) was absolutely right then, and it still holds up now, that "to write, to engage in any communication, is to participate in a community; to write well," what I think of as verisimilitude, "is to understand the conditions of one's own participation" (p. 617). I believe Miller's argument made possible current conversations about social justice. Narrative is what helped me envision the humanistic pedagogical design that later became my dissertation (Bridgeford, 2002). I've described narrative's role in my technical communication courses as "narrative ways of knowing" (Bridgeford, 2002, 2004) and as a process of enculturation used by communities of practice (Bridgeford, 2007).

To illustrate those concepts, I turned to *Terrarium*, a dystopian novel, as a context, for all the reasons Barton and Barton (1988) extoled about the virtue of narratives: they are "read faster," "processed more effectively," and "remembered better" (p. 43). In much the same way, I have described

Terrarium to students and colleagues as a quick read, as an interpretable common context, and as the narrative accrual (institutional memory; 2004) for action. Additionally, I was influenced by educational psychologist Jerome Bruner's (1991) "The Narrative Construction of Reality" because he reflects what we believe about the humanistic perspective we have come to rely on in the field. He says that narrative, and the culture from which it grows, "is never 'point-of-viewless'" (p. 3). Natasha Jones (2016a) emphasizes the same problem with our dominant narratives when she says that we must "understand that technical communication is not neutral or objective," but it is, rather, "political and imbued with values" (p. 345), that is, embedded in a context. Without this focus, she concludes, our field might not be as "humanistic" as we'd like to believe. Whatever the story (fiction or technical document), narrative is what helps us keep our humanity in focus and makes us accountable to an audience that will know if our efforts are true.

Terrarium in the Classroom

Although I have tried other narrative contexts (e.g., *Dune*), I have mostly used Scott Russell Sanders's *Terrarium* (1985) in my classroom. The story of *Terrarium* is one of compliance and defiance. As I have indicated, this dystopian novel depicts a totalitarian society that resulted from an environmental problem in which the earth has become so polluted that to survive, humanity was moved into large, dome-like structures called enclosures. The current timeline of the story is roughly 20 years since humanity was moved inside (2026). Residents dress in the same kind of clothes, paint their faces in the same way, and, perhaps most important, share the same fear of the "Wilds." Very few people in this timeline remember the Wilds and are sufficiently afraid of "Terra" that they willfully conform to the totalitarian power structures that define life in the enclosure. In the novel, citizens add color to their lives by wearing mood gowns, elaborate clown makeup, and outlandish wigs. They ride pedbelts to electroball courts, participate in rhetoric matches, and visit Disneys (museums) and eros parlors. There is no mention of politics because no one is allowed to speak out against what I call the Enclosure Board or to leave the enclosure. Residents are the same only in that they are all forced to follow the same strict rules of the enclosure society designed to keep them safe from pollution. I use *Terrarium* in particular because by depicting a futuristic society, the story has technical devices with which students are not necessarily familiar (e.g., vidphone, vidscreen, vaporizer, etc.); at least that was true when *Terrarium* was published in 1985. So, my first step in rethinking my narrative pedagogy was to decide whether this novel could engage social-justice issues given its totalitarian society. In fact, the restricted societal rules offer, perhaps, a perfect context for situating students in ways that enable them to use tactics.

To contextualize the assignment scenarios discussion that follows, it will help the reader to know more about the plot of the novel. The action of the story focuses on the goal of "the ingathers," who want to escape the enclosure (that "monster of rationality," Sanders, 1985, p. 13) to live in harmony with nature. The plot follows Phoenix's transformation from an obedient enclosure resident to a Wilds enthusiast, which started because of his sexual attraction to Teeg. As I see it, this plot supports an assumption we often make when confronted with a societal problem—it's only an ethical dilemma when the problem is monstrous and therefore distanced from our everyday lives. If we focus only on monstrous situations, it will be easy to ignore the small, mundane ways that affect our lives every day, such as the microaggressions intended to position other people in ways intended to silence them. Practicing technical communicators face difficult decisions every day when creating documentation (see Loftin & Cryer, 2015, pp. 25–26 for a fuller discussion). Ethical dilemmas occur every day in our lives and shape who we are even if we can't see it.

I have become increasingly aware of just how much the scenarios I have created based on *Terrarium* are just providing too easy of a path from which students, as well as myself, could hide behind our positionality and privilege. The original and revised scenario that follows focuses only on the scenario and not the verbiage for the specific assignment expectations such as length, illustrations, sources, and so on. After providing the scenario, I generally include specific assignment instructions such as, "In groups of three, write a recommendation report that addresses the Enclosure Board's concerns. This report should be at least 2,000 to 3,000 words and include illustrations that support your recommendations." For most assignments, I don't allow students to use existing characters because they tend to focus only on what's in the novel instead of imagining ways to act from a weak position. Typically, I provide a list of possible areas or devices to get them started, depending on the project. Although I provide opportunities for students to invent a topic of their own, they tend to pick one from the list I provide. Let me highlight an original assignment scenario that demonstrates, I think, a weakness for situating students within agentive action:

> Because most documents before the Enclosure Act of 2026 were lost during the dismantling, Enclosure researchers with the newly established Historical Research Society (HRS), which is part of the Institute of Global Design, have been charged with the job of chronicling historical precedents of the Enclosure Act. The researchers are especially interested in ecological problems and how arguments were interpreted and acted upon. For example, climate change has been argued to be the result of human activity. These research documents will be collected, cataloged, and presented at the unveiling of the new Environmental Awareness Exhibit of the Historical Museum of Ecology. As technical communicators in Zuni Franklin's office, you have been asked to help with this research by exploring the various perspectives affected by environmental actions: legislative, judicial, political, economic, social, religious, cultural, or historical. Choose one of these perspectives and write a 500-word report detailing an historical problem as it relates to an ecological problem and the actions that led to it. For example, you might research the damage homesteading practices had on the ecology of the prairies that eventually led to the dust storms of the 1930s affecting the soil's ability to store moisture. The reports will be used to inform the creation of museum displays.

It did not work well, especially in an online environment. I realized that by using the novel's current timeline and creating the Historical Museum of Ecology for the Environmental Awareness Exhibit, I was situating students as citizens within the enclosure in ways that required them to comply with the rules of that society with no allowable resistance. At the time, I assumed that students would just gravitate toward a variety of groups to focus on for the exhibit, but they, not surprisingly, focused on people most like themselves. Further, by situating students as technical writers in Zuni Franklin's office, I was asking them to comply with the rules of the enclosure society because Zuni, one of the architects of the enclosure, also serves as what would be equivalent to the CEO of the Institute of Global Design (IGD), an organization featured prominently in the novel. Even though we later discover that Zuni was strategically maneuvering her eventual escape, she did so in ways that didn't hurt the enclosure and its operations because she believed she was protecting the earth by keeping people inside.

The scenario obviously falls short as a narrative construction of social justice because it does not situate students within a problem-solving situation, ask them to be critical thinkers, or help them look beyond the enclosure society described in the novel. Therefore, students tended to write to an audience of ubiquitous enclosure residents, instead of specific groups with little or no power. The assignment resulted in content that would be acceptable only within the power structures of *Terrarium* with little imagination needed. In fact, I constructed the scenario as if a single enclosure held people of only one type of community member. Given that each of the 130 enclosures holds about a million people, it stands to reason that when moving people inside, the enclosure communities would have included people of all races and ethnicities. I was drawing the scenario as well as my expectations for performance with the same brush strokes. Clearly, I wasn't asking the right questions.

Revised Scenario

> It's been about 50 years since the Enclosure Act of 2026. Researchers with the Enclosure Board have been asking enclosure leaders throughout Oregon City to serve on the newly established Community Board for Diversity, Equity, and Inclusion (DEI). Before making any decisions, the board would like to understand how some communities have been disenfranchised. As freelance technical writers, you have been hired to conduct background research on one of the communities listed below in relationship to an area such as housing, finance, employment, recreation, safety, entertainment, health, or education. Background research should characterize the community, provide a description of the community's history in the enclosure, contextualize the disenfranchisement, and make recommendations for improving its conditions. Submit your recommendation report to the DEI committee chair, Adella Smith.
>
> Communities (choose one):
>
> - Immigrants from a non-English-speaking enclosure outside the United States
> - Parents

- Credcard administrators
- Enclosure teachers
- Health patrollers
- Rhetorical matches managers
- Eros parlor guests
- IGD employees
- Master troubleshooters
- Shasta Game Park operators
- Other (with approval)

With the revised scenario, my goal is to situate students in ways that require them not only to create technical content for a specific audience located in Oregon City, but also to situate them within a specific problem that in some way mirrors recent social-justice issues, thus enabling them to bring an issue into focus. The supposed homogenous nature of the society in *Terrarium* is the result of author intention to create a society in which the typical enclosure resident is the same as any other resident. Given the current timeframe of the novel (circa 2042), these rules were established by the strategic power brokers who likely continued the systemic racism that existed before moving humanity into the enclosures. In the revised scenario, I put the timeline farther into the future than the novel's time frame to encourage students to think beyond what's only in the novel as well as to make the society depicted more accommodating to difference while still allowing them perhaps to pull from current problems for their research. The revised scenario helps situate students in ways hopefully out of their comfort zone, as well as my own, by asking them to act in ways that address issues related to social justice. The revised scenario example highlights not so much a cause that might or might not be theirs but also a complete decision-making process that helps them address a social problem in real, albeit imaginative, ways. Students who successfully poach in the *Terrarium* world do so by drawing not only from what they are learning about technical communication but also from their own belief systems. How they choose to act must be their choice and reflecting on that choice should be part of the assignment.

By centering my pedagogy around narrative, I connect students to a sense of humanity on dual universal and individual levels. It also helps that, as Bruner (1991) says, narrative is one of the ways we "work 'mentally' in common" (p. 20). Students engage the narrative in ways that enable them to move beyond the "efficiency-driven narrative" that has dominated our field (Jones, Moore, & Walton, 2016, p. 281). Much like Miller's (1979) discussion of positivism's influence on technical communication pedagogy, Bruner similarly notes that "unlike the constructions generated by logical and scientific procedures that can be weeded out by falsification, narrative constructions can only achieve verisimilitude" (p. 4), or what Miller describes as a "persuasive version of experience" (p. 616), that is, a plausible presentation of a truth. In writing technical content, students must achieve verisimilitude through essential skills such as interpretation, analysis, style, structure, and design. Given *Terrarium*'s story about a totalitarian society, narrative works, as both a strategy and a tactic—the power structures of master narratives and the tactics individuals use to sidestep them (de Certeau, 1984). Whatever technological narrative we create, we experience that moment of eureka—verisimilitude—when how to act, what process(es) to use, and what to write become a reality. Students who experience that reality, I think, are using tactical reasoning.

In reassessing my pedagogical habits, I found a welcome opportunity to reflect on my motives and to question their outcomes especially when the events and conversations that emerged more prominently with George Floyd's death and the discussions that followed challenged me to confront my own positionality toward and history with social justice. Soon after Mr. Floyd's death, I hungrily began reading articles, chapters, and books about social justice and technical communication beginning with Angela Haas's (2016) article, "Race, Rhetoric, and Technology: A Case Study of Decolonial Technical Communication Theory, Methodology, and Pedagogy," and Rebecca Walton, Kristin Moore, and Natasha Jones's (2019) *Technical Communication and the Social Justice Turn: Building Coalitions for Action.* From these events and readings (for more discussion on decolonizing technical communication, see Agboka, 2014 & 2020), I was encouraged, nay convinced, that I needed to rethink my pedagogy by considering aspects of myself I had not necessarily faced concerning social justice. In this, what I consider my own tactical narrative, my positionality revealed itself in ways too strong to ignore as I examine how I did it.

Writing this chapter has helped me face some uncomfortable truths about myself, including a lack of awareness and action given my positionality, privilege, and power. In *Technical Communication after the Social Justice Turn*, Walton, Moore, and Jones (2019) point out that "recognizing these truths can render us afraid to act" (p. 5), which describes my situation. My pedagogy was not as transparent as I believed it was simply by virtue of being a narrative. After reading Walton, Moore, and Jones, I realized that I was not achieving all the pedagogical goals I thought I was. I was afraid to act, to be explicit, to teach more purposefully, and to address social justice directly and without fear. So, I might be late to the conversation, but I'm here now.

Positionality, Privilege, and Power: A Self-Reflexive Exercise

My administrative positions as director of our graduate certificate in technical communication, as internship director, as graduate program director, and now as department chair provide me with a level of authority that enables certain privileges, that is, power structures, such as acting as the gateway into our graduate program or graduate certificate or writing a vision for the department that I call "English Without Borders." Until last year, before the pandemic, I didn't think I had any authority to participate in what I had thought were issues unrelated to me. I used to believe I had to have certain credentials to talk about issues related to social justice, issues that seemingly affected me only in an adjacent theorical exercise. But, as I discovered, and continue to discover, not only can I have a voice in these discussions; I can also do the work of social justice. In the last chapter of *Technical Communication After the Social Justice Turn*, Walton, Moore, and Jones (2019) "address the critiques and questions that emerge for readers and skeptics" (p. 157). In particular, these authors identify thoughts and critiques that people use to keep themselves from getting involved. For me, I identified mostly with the sentiment that "[i]t's so daunting that injustice is institutionalized and systemic. What can I—just one person—even do about it?" (p. 163). I even thought, "Who am I to do anything?" I find it uncomfortable to realize that some of what these authors address can be applied to me. I felt powerless to act or found it easier to go along with the existing power structures.

But then a colleague from the Department of Black Studies asked me to cohost an event we called "What UNO Do You Attend?," an event meant to bring issues of racism at my school to the surface. At the event, I listened primarily to students of color tell powerful stories about the microaggressions they have endured at my school, seizing an opportunity perhaps to affect those structures. I saw what my colleague meant, and now I cannot unsee it. I had hoped that more white students would attend but was not entirely surprised that they didn't. This lack of attendance, I now see clearly, reveals the belief that racism is not a white issue. I was also in a position to do something. We had planned a follow-up event, but then the pandemic pushed us all inside. Earlier the same semester, I attended a session on microaggressions, the lessons from which are unavoidable, as they should have been. Thus began a shift in my thinking and a transformation that continues to inform my thinking. However, this is a slower process than I thought because I looked at my approach to the capstone course in technical communication that I taught fall of 2020 as highly evolved because I focused on social justice. In hindsight, I think I was guilty of "labeling [my teaching as] . . . 'social justice' in service of one project or one class without fully committing to a vision of the field that is inclusive and just" (Walton, Moore, & Jones, 2019, p. 4). In addition to a variety of articles and chapters, including *Technical Communication and the Social Justice Turn*, I did require students to work on service-learning projects. But I wonder now if I was truly engaged in it or if "I employ[ed] a charity model that fail[ed] to redress inequities" (Walton, Moore, & Jones, 2019, p. 4). If I was, in fact, employing a "charity model," then I have more rethinking to do, to say the least. Based on students' end-of-semester reflections, they loved the topic, enjoyed discussing it, and wanted more. That's a good start.

Relatedly, I had a discussion with a Political Science Department colleague who asked about my approach to teaching technical communication. She was interested in how I used a novel as a context for technical communication in particular, asking about my selection process. I explained that my novel of choice, *Terrarium*, works well in technical communication courses because it depicts a technological problem and because, as a dystopian novel, it typically includes a nomenclature useful for writing definitions, descriptions, and instructions. I told her that *Terrarium*, an Earth Day novel in 1985, depicts a story about the dangers of pollution and its effects on the earth, providing ample room for me to focus the class on a variety of issues. In the story, the earth has become so polluted, they were forced to move humanity into large dome-like "enclosures" (called Oregon City, China City, India City, and so on) that were built on land or on sea. The action takes place 20 years or so after Enclosure Day (2026), when humanity was forced to move inside the enclosures. As an aside, I added that students identified with the characters because they are primarily white, as are my students. She asked, "How do you know they're white?" Indeed.

Several aspects of my positionality, privilege, and power peeled away in that moment. I surmised that most of the characters were white, I'd like to say, because the author deliberately spends significant time describing a group of six characters who represent, he says, a "rainbow of flesh" (Sanders, 1985, p. 63). Of the six, the two main characters are white, and it is their love story that moves the plot forward. But it's probably truer to say that I drew this conclusion because the author spends no time creating identity markers for most of the characters, and I assumed they were white because that matched my own identity markers. In *Terrarium*, that is, inside the enclosures, reference to anything physical about the body was considered taboo, symbolizing how far removed from natural environments the characters are situated. The characters dress in "mood gowns," apply heavy clown makeup or don masks, and wear elaborate, colorful wigs. No physical aspects of the body are exposed. Given these hidden identity markers, why then did I jump to the conclusion that they were white? Obviously, because I was seeing only people who looked like me. I also realized that for the past 20 years, I have had only two persons of color in all my classes combined, a demographic that needs to be addressed in my program. So, not only was I painting *Terrarium* characters with one brush, I was also surrounded mostly by students who looked like me, something I didn't consciously notice until last year.

Technical Communication Pedagogy and Social Justice

In a blog post following a 2015 ATTW panel presentation, Kristin Moore (2020) says that she and her colleagues sought to "articulate the value of technical communication as a site of social justice." When reflecting on this topic, she says that the panel participants were disappointed because those in attendance were "still asking" how to incorporate social justice into their classes and existing pedagogical frameworks. I wasn't able to attend ATTW that year, but if I had, I'm sure I would have been at that panel. I likely would have been one of those present who was "still asking"; in fact, I might very well have been asking for the first time. Although I don't agree that we should ever stop asking how, something I see as speaking from a weaker position, I do agree that there are many "institutional and disciplinary difficulties of enacting social justice in technical communication" that hinder local efforts to decolonize pedagogy and perhaps come closer to the humanistic framework we desire. And Moore is absolutely right when she says that "we need more programs and textbooks to support the social justice objectives." But I now know that we have been asking how and for quite a while. In 2013, Kristin Moore stated that "we have yet to develop pedagogical tools to help students adopt strategies that respond to diverse situations" (p. 66). Our academic lives often get in the way, especially if we take administrative posts. More importantly, I don't think we are always ready to hear and accept an argument as my previous discussion shows. I heard someone say once that tenured scholars grow comfortable with the theories they read and the pedagogies they practice likely created in graduate school and solidified during the tenure-track years. I'm not saying that this is a justifiable position, but I do understand it; the energy it takes to change a mindset is exhausting, as well as efforts to theorize technical communication are relatively young/emergent, which makes us cling even more fiercely to the relatively few tools in our field's theoretical tool kit. By not changing a mindset cultivated over 20 years, I (we?) am letting myself off the hook not only by being unwilling to change but also perhaps by not willing to understand how to change.

Although a focus on "how" to bring social justice into the technical communication classroom offers an opportunity for action, there have been pedagogical discussions specifically about social justice for at least a decade, some of which have significantly influenced my thinking. Angela Haas's (2012) case study of her development of a class on race, rhetoric, and technology using a "decolonial" framework led me to "challenge myself," just as she had done, because it is important to "consider how the language we use might serve to redress the long-standing legacies of colonialism and imperialism" (p. 288). For me, that language has always fit within a humanistic framework. My scholarly journey began with Carolyn Miller's (1979) humanistic rationale, opening the scholarly door to successive pedagogical discussions including social construction (e.g., Blyler & Thralls, 1993), ethics (e.g., Dombrowski, 1999; Dragga, 1999), narrative (e.g., Barton & Barton, 1988; Byler, 1995), political (Blyler, 1998), cultural studies (e.g., Longo, 1998; Scott, Longo, & Wills, 2006), and diversity (Jones, Savage, & Yu, 2014; Savage & Mateeva, 2011; Savage & Mattson, 2011). Despite this history, we have further to go. Natasha N. Jones's (2016a) argument about the humanistic perspective led me to confront my mindset and to question whether our field is indeed humanistic. If it is, we need to show it better, as Natasha N. Jones (2016a) argues, by more "directly engag[ing] with issues of injustice, inequality, and dehumanizing" in ways that truly focuses on "human experience"—all human experience (p. 347). In their discussion about how technical communicators can enact social justice, Jared S. Colton and Steve Holmes (2018) describe how it's possible to change from a "passive equality" that waits for "institutional sanction" to "active equality," which better reflects a more ethical social arrangement" and translates into a more "active practice" through "greater critical attention to social justice" in our classrooms (p. 13). This is just a sampling of a few of the articles I look to for guidance when thinking about and including social justice in my pedagogy through narrative. I also think it's worth acknowledging the potential for harm as privileged, predominantly white scholars consider pedagogical issues related to race and social justice. We can get it wrong and hurt people in the process. This risk reinforces the need for open conversation and collaboration surrounding the topic, a deliberative process that this chapter invites by sharing my own social-justice journey.

Narrative and Technical Communication Pedagogy

In the *Narrative and Professional Communication* collection, Perkins and Blyler (1999) identify the "narrative turn," encouraging "teachers and trainers to refocus their pedagogy on professional communication as knowledge-making," a perspective with the potential to "revitalize classroom and workplace teaching" (p. 18). Nancy Blyler (1998) explains that for her, the "narrative turn refers to more than a discourse form to be included in manuals, proposals, and reports. Rather, narrative is also integral to the way we understand our existence and construct ourselves as human beings" (p. 2; see also Blyler, 1995). From this perspective, narrative is clearly a direct path to our scholarly and pedagogical claims of a humanistic perspective. Several scholars in the 1990s did focus on the value of storytelling in scientific and technical contexts (Journet, 1999; Herrick, 1999; Smart, 1999), which is returning as a valid conversation (an upcoming special issue of *Technical Communication* on storytelling and previous issue of *Intercom* on technical storytelling). In the client-based project she assigns, Kristin Moore (2013) focuses on public policy and the telling of stories demonstrating ways students can "build relationships with citizens [that help students better] understand [how stakeholders] are connected to the place/city/landscape, the policy, and the other people involved" (p. 64). In some ways, the narrative context I provide through *Terrarium* enables the same kind of activity—students read, interpret, understand, and reflect on characters' actions in ways that help them achieve a dual empathy that enables them to learn about their audience as well as themselves.

Although I continue to use a novel as a context, my goal is to shake up that pedagogy by decolonizing it, beginning with a rethinking of the narrative context I use for assignments followed by an analysis and revision of my assignment scenarios. Acceptance of the narrative turn, albeit not immediate, has gained some important ground in more current conversations. Natasha N. Jones's (2016b) discussion of narrative inquiry and a human-centered research design "encourages reflexivity" and enables the designer to take "into account his or her own positionality" and to "acknowledge . . . worldview, opinions, biases, and stance" (p. 482). In the same year, Jones, Moore, and Walton (2016) apply the notion of antenarrative, a "disruptive 'before' story that seeks to destabilize and unravel aspects of the tightly woven dominant narrative" (p. 40), which enables a disruption to the original narrative that has dominated. Creating disruption, Jones (2016a) says in another article, enables us to reveal "implicit master narratives we hold in ways that "take apart the story, revealing underlying texts and giving voice to things that are often known intuitively" (Jones, 2016a, p. 350). As I go back and retrace the history of my pedagogical thinking, that is, create my own antenarrative, I have been actively dismantling the technological narrative (Kimball, 2006) I own by using a tactical lens that has helped me circumvent it.

Difference in the world of *Terrarium* is described according to the clown makeup, wigs, and mood gowns; perhaps this was Sanders's way to handle difference. But that kind of difference is fantastical and might not be relatable to students and the differences that exist in the world today that technical communicators need to be aware of as they pick the appropriate language. The only time Sanders mentions diversity outside the clown-like appearances is when he describes the members of the Ingathers who escape the enclosure, calling them a "rainbow of humanity." But nothing about the original scenario brings students out of their comfort zones in ways that enable them to act in a tactical way, nor does it require much from me as an evaluator. This assignment provides the potential to engage in explorations of diversity, equity, and inclusion in ways that shed light on one's (and others') position, privilege, and power, operating "outside, between, and through" existing power structures (de Certeau, 1984). Narrative, according to Jones and Walton (2018) is "well suited to this complex work of exploring positing and revealing privilege because of the dialogic space it opens up" (p. 251). Narrative encourages, they argue, "ethical reasoning and exploring positioning and revealing privilege," a goal, perhaps, for all our pedagogies that support a "humanistic" discipline (p. 247). For example, one scenario I wrote asked students to write a policy for reporting infractions to the Health Patrol. Later, I asked them to revise that policy because the rules for the enclosure were becoming less oppressed. Another scenario situated students as a member of a community-wide housing committee that was charged with revising policies established upon moving into the enclosure that restricted the housing units based on race. Not only would students have to interpret the context of the novel, but they would also have to impose differences within its story to revise the policy. In this way, narrative works effectively as a context for "critical thinking about social justice," as Jones and Walton (2018) argue, because it has the capacity to help us expose the power structures that limit action (p. 243), and, I would argue, to imagine and accept different ways of knowing.

Upon reflection, I realize now that while I thought I was being explicit with students as to the role of narrative, in hindsight, I left too much unsaid. Although I did recognize it as a tactic (a "textual way of operating," de Certeau, 1984) that helps us as instructors engage students on a human level, I thought the moral component of narrative was obvious. I explained to students how to enter the story imaginatively and engage in the action, but I did not explicitly talk about narrative as a means for understanding the human condition. My thinking at the time was that narrative inherently provides a moral context, and therefore, whatever technical content students create based on their interpretations automatically put them within an ethical context because narrative must be interpreted to have meaning, as Jerome Bruner (1991) says. Students didn't seem to have problems creating documents or thinking of *Terrarium* as a context. It wasn't until I started teaching asynchronously that I realized it wasn't as obvious as I thought. In person, it was easy to tell whether students were confused or elaborate on an assignment in the moment. I made movie lectures for the online environment, but they didn't seem to have the same clarity offered in synchronous conversations, a common problem. But, while previous students seem to understand my intentions better, they still created content within the power structures of a totalitarian society that restricted their agency. The creativity of their content swayed me. Using *Terrarium* as a context, for example, one student group created what they called "Mate Crime," which referred to the 12-step permissible mating process that begins with eye contact and moves slowly through increasingly progressive phases, eventually engaging in total intimacy. The students created Mate Crime, but within the confines of the enclosure society, which gave agency to the policing entity, the Health Patrol, not the students creating the content. It was creative and did create verisimilitude, I think, but it just wasn't a tactic.

This context makes available the capacity for action within the story's strategic power structure. de Certeau (1984) defines strategies as systems of organized power, and *tactics*, specifically *la perruque*, as the creative appropriation of everyday practices that he calls "ways of operating" or more specific to my discussion, "textual ways of operating." Textual ways of operating—or what de Certeau refers to as narration—are "an art of speaking" that "exercises precisely that art of operating" (p. 77). In other words, to participate in an everyday practice one has to be able to also engage in language about the practice. This language creates a "fictional space" that "makes a *coup*"—creating a way of "exercising itself than by the thing it indicates" (p. 79; italics in original). Earlier in the preface, de Certeau (1984) uses reading as a way to demonstrate what happens when we read: the reader, he says, "insinuates into another person's text the ruses of pleasure and appropriation": he "poaches" on it, is transported into it," "[slipping] into the author's place" (p. xxi)—what I see as an act of imagination that exercises itself. In this way, I ask students to slip into *Terrarium*'s story in order to act, to engage in textual ways of operating. How students insinuate themselves determines whether they have made a coup through the invention of technical content.

To ask students to engage in textual ways of operating, I create an assignment sequence that includes situating the novel as the fictional space for tactical action. I see this kind of agency operating as a tactic on my part as well as on the part of students. For myself, I need to provide the context in which I can construct ways to provide students within the capacity to act, to poach in the novel's totalitarian regime, and affect change. For students, this kind of agency enables them to act out a given scenario within the context of the novel through their imaginations. I have called this activity "imagination as agency" (Bridgeford, 2019), a schema that enables members to recognize what constitutes experience in a community of practice for themselves as well as others. For example, when asking students to write letters of application to Dr. Passio in the novel, one student addressed Dr. Passio's dream of Project Transcendence, a method for freeing people completely from Terra by creating enclosures meant for space. In his letter, one student insinuated himself into this little-discussed aspect of the novel, extending it to how he could help Dr. Passio realize that dream through his own idea of survival in space. The student effectively created a sense of belonging that Etienne Wenger (1998) calls an "identity of participation"—his notion of agency, which depends on whether community members can see themselves and others acting in situations long before actual action takes place. I see imagination as an agentive lens that can help teach students how implicit knowledge works; the novel provides a textual doorway for students to be able to see themselves as part of the story and, hopefully, imagine ways for acting that poach on and affect the story. They need to decide how they will act in order to create the technical content assigned, moving fictional details around, changing details, adding or subtracting from the story, creating characters, and so on. However, I have come to realize that I wasn't so much opening a space for students to exercise agency as I was situating students in ways that removed their agency as well as their ability to act tactically.

Journey of Transformation to Social Justice

In the beginning, I picked *Terrarium* because as a dystopian novel, it offers nomenclature ready made for writing definitions, descriptions, and instructions. But in all honesty, I also picked it because it is easy to read, easy to interpret, and easy to identify meaning. By easy, I mean that I didn't have to look beyond my own positionality. The ubiquitous enclosure residents in the novel have no identity markers. They dress in clown makeup, mood gowns, wild wigs, all of which hide all their identity markers. In fact, when Phoenix sees Teeg, the "barefooted woman," for the first time, she is walking backwards on the pedbelt (moving sidewalk), a physical act, which is odd enough to him. But she also appears al natural, revealing her real hair, her real face, her bare feet, and the shape of her body. When she refers to his "flab," Phoenix thinks, "[H]ow dare she refer to [my] body!" A few minutes later, he refers to her feet as her "walking things" because he was "unable to bring himself to name a body part." He compares his clown-like appearance to hers, noting that his body is properly hidden, "every inch of flesh cloaked in a mood gown" (Sanders, 1985, pp. 4–5). Of course, the wearing of mood gowns, makeup, wigs, and hoods reflects one of the novel's themes, that is, the distance between humanity from Terra, the earth, and the danger that it invokes. This distancing makes Phoenix's come-to-nature transformation all the more meaningful. This distancing from even the smallest identity marker is typical of dystopians like *Terrarium*. Issues of diversity are easily hidden in the dominant narrative. Everyone can look the same if there are no visible forms of difference. The reality is that issues of diversity are hidden within dominant narratives.

The banality of the novel enabled me, as well, to distance myself from having to reveal or bring attention to the characters or my own identity markers and their connotations. My pedagogical narrative is one of tradition and convention. But my "antenarrative" is the journey to a narrative construction of social justice. From a traditional narrative perspective, I would assign students to write technical documents to the ubiquitous enclosure residents. For example, I asked students to write instructions to enclosure residents for one of the new technical gadgets (e.g., vidphone, vidscreen, vaporizer, etc.) in the process of moving into the enclosure. These instructions would be available to everyone in the enclosure through the cyberboard (internet). I was in effect asking students to write to everyone, which we know doesn't work. With a couple of assignments, for example, I provided this scenario:

> The year is 2026, and people are about to move into the enclosure. Because the environment and tools will be different from what people are familiar with, the Technical Writing Division that you work for has been asked to provide various rhetorical elements to include on the enclosure cyberboard, a content management system (CMS). This system is accessible by anyone from anywhere in Oregon City.

I then asked students to "write an extended definition for a term" or a "process or product description for one of the glossary terms . . . or a term from your own imagination." I provided the list of terms directly from *Terrarium*, and no one ever elected to use their own imagination. Even when I tried to differentiate residents by grouping them according to geographical region, as another example, I still stayed within the traditional narrative approach:

> Although the enclosure system seems characteristically homogenous, many cultural differences developed within individual enclosures. With over 100 land and 30 water cities in the United States alone, single enclosures have evolved into individual cultures defined by their legacy customs. Over the years since the Enclosure Act of 2026, these communities have, not surprisingly, developed the historical traits characteristic of their geographical regions. Nebraska City, for example, has shown both the strong agricultural traditions when humanity lived and worked on the land and the more modern telecommunications industry that defined its place in the early part of the twenty-first century. Given the cultural differences, students participating in the academic exchange program, particularly those from Oregon City, will need guidance with the technical aspects of their room and board as

> well as the cultural customs specific to Nebraska City and the Midwest Enclosure University. The board is asking for a report about the technical, cultural, or social process associated with Nebraska City. Drawing from *Terrarium*, groups should extrapolate information about enclosure living from their knowledge of Oregon City and make it relevant to Nebraska City.

As creative as I thought I was, the assignment still characterized enclosure residents ubiquitously using one set of identity markers. There was no antenarrative in this or any of the other *Terrarium* assignments I've created over 20+ years. Although I ask students to consider the meaning of *Terrarium* as a means for providing the context, something missing is that I don't ask students to critique any of the core premises in *Terrarium* such as how to identify issues of power and the instances that stand in the way, or why there are no identity markers and what it means when the author does identify them (rarely), and so on. I was fearful that the course would turn into a literature course, which was feared at a certain time in the field (e.g., Allen, 1989).

I see a narrative construction of social justice as the establishment of a narrative context from which teachers and students can act with agency and empathy, giving "voice to the marginalized" (Jones, 2016a, p. 350)—as a tactile action. Upon finishing the novel, I do ask students to engage in a discussion about the meaning of the novel, that is, what's going on, their interpretations of its meaning, and the actions it affords. For example, while teaching at Michigan Tech, I triangulated my use of *Terrarium*, technical communication, and Torch Lake, an identified Area of Concern in Michigan by the Environment Protection Agency. Using the concept of exigency, I asked students to define the exigency of *Terrarium*. One student wrote: "The exigency of *Terrarium* is Torch Lake." This statement is the whole reason I continue to use *Terrarium*. But I didn't explore it further with the entire class, missing an opening for keener insight into difference such as having fresh, clean water to drink and technical communication's influence in that context. The assignment scenarios then draw students into both the story itself and the reality I invent through scenarios, providing opportunities for them to adjust, to interpret, and to act, that is, using *la perruque* tactics from which they can establish a resistance to the established power structures. It took me a while to learn how to write an effective scenario that actually used tactics, but I still need more practice.

When I first taught technical writing as a graduate student, I think I was more creative, albeit in a limited way. As an in-class exercise, I asked students to read an excerpt from Wil Weaver's (1990) short story "A Gravestone Made of Wheat" and write definitions. Briefly, this story focuses on a farmer's wish to bury his wife on their farm but is informed by the sheriff that because of public-health ordnances in North Dakota, it is unlawful to bury on private land. In the midst of this dilemma, the farmer reminisces about when his mail-order bride arrived from Germany in June 1918; they were denied a marriage license because she was German, and the marriage would give her citizenship. Throughout the rest of their lives, they behaved as if they were married. I gave students a list of terms from the story (e.g., *farm, marriage*, etc.) and asked them to write a definition for each from the perspective of one of the characters. In his reflection of the assignment, one student wrote, "It never occurred to me that the definition would be different just because he was a judge." Shortly after, I read *Terrarium* for the first time, and the rest is history. I missed a prime opportunity to focus on audience and/or immigrant issues in technical communication.

By transforming my narrative framework into a personal tactical pedagogy, I continue to work to establish a context for action through narrative, introduce or highlight a social-justice problem into that narrative (if it doesn't already exist), and enable agentive participation through interpretation and reflection—each of which enables human mindfulness. It creates opportunities for pedagogical action that demonstrate for students the outcomes of my course. I am working to find more and better ways to question my own positionality, privilege, and power while hopefully leading students to do the same. The tactical framework I create provides a story of the human condition and with all its social problems, showing students examples of both good and bad participation (as depicted in the story), enticing them to engage in professional development. They then participate in ways that demonstrate an awareness of their own positionality, privilege, and power.

With this tactical reflection, I am beginning to question how my identity affects others, how the authority of my identity provides different opportunities for action (both from and to me), and how the power differential of my identity affects meaning, or not, as the case might be. I constantly ask myself, "Can I do that, and, if so, can I do it effectively?" Perhaps not comfortably at first, maybe not ever. The revised scenario I mentioned earlier, for example, pushes me out of my comfort zone in ways that reveal to me real attention to "what my identity means in particular contexts of action" (Walton et al., 2019, p. 63). As I mentioned earlier, my identity markers show that I am a woman, white, tenured, and full professor, as well as a department chair, with all the privileges and powers those positions entail—all of which have "dominant ways of knowing and dominant notions of credibility" (p. 78). Even if students don't know that I am the department chair, being the teacher at the front of the room is enough to demonstrate a powerful identity marker. At the same time, I am currently an imagined version of myself trying to follow a more coalition path by addressing social-justice issues directly in ways that I'm not comfortable voicing.

For one thing, creating tactical narrative assignments offers the potential for imagined social-justice action in which students can then participate. When students begin researching the communities in the scenario in relation to an area such as housing or safety, for example, they might choose to research current migrant issues at the borders or red-lining practices in a particular city and apply what they learn to the context and residents of *Terrarium* (or another narrative context), perhaps engaging in dual empathy. In turn, students develop professionally by learning to identify boundaries for their actions, to imagine their participation toward that social justice problem, and to act on the problem. Within the safety of the classroom and the context of the novel, students practice conducting themselves within a community, which helps them imagine their participation in ways that break down the confines of the classroom. Students write technical documents more aware of audience, situation, and purpose. To act, they must interpret the context to determine their participation. Imagining that action enables them to reflect on their participation and that of others—both good and bad. I call this activity imagined tactical participation. The act of researching and writing the assignment as well as its accompanied reflection occur in my and students' imaginations, participating as we and others would do.

Conclusion

I think it is safe to say that my use of narrative in technical communication courses frames my pedagogy as humanistic because, as stories about human beings and human events, narrative inherently "relates to what is morally valued, morally appropriate, or morally uncertain" (Bruner, 1990, p. 50). I see stories as already morally relevant because "to tell a story is inescapably to take a moral stance, even when it is a moral stance against moral stances" (Bruner, 1990, p. 51). Until I sat down to rethink my pedagogy, the moral aspects of narrative seemed self-evident, but that doesn't mean students automatically recognize what is, or what I (or what they) consider to be, ethical or unethical behavior. And even if students do, they might not have reflected on it in actual practice. As students navigate through the reading-and-writing-assignment sequence, they take away from it, perhaps, a clearer understanding of our humanistic discipline. Natasha N. Jones (2016a) asks, for example, what can we do to "move forward to further legitimize our field and empower scholars" while also "[valuing and legitimizing] other perspectives and experiences?" (p. 345). She says firmly that we need to "directly engage with issues of injustice, inequality, and dehumanizing forces" (p. 347). And, in doing so, we might be more willing to act outside the strategic confines of institutional power and think more tactically about our pedagogies.

In writing this chapter, I tried to do just that. From de Certeau (1984), I learned to think tactically in ways that I can pass on to students. Of course, one chapter can't provide all the answers instructors need to build a narrative construction of social justice in technical communication pedagogy. It's important that we continue to focus on the humanistic aspects of our field especially in ways that help us decolonize our pedagogies, and this is something we can work on right now and in the future. In other words, it is increasingly important to question, evaluate, and address how we have made determinations about other ways of seeing. We must continually seek out pedagogical moments that help us reflect on our own teaching, which I see as a career-long challenge. I see the social-justice pedagogy described in this chapter as only a glimpse into what I can/should do. For me, it is a step toward a journey of instructional self-discovery.

References

Agboka, G. Y. (2014). Decolonial methodologies: Social justice perspectives in intercultural technical communication research. *Journal of Technical Writing and Communication, 44*(3), 297–327. https://doi.org/10.2190/TW.44.3.e

Agboka, G. Y. (2020). "Subjects" in and of research: Decolonizing oppressive rhetorical practices in technical communication research. *Journal of Technical Writing and Communication, 51*(2), 159–174. https://doi.org/10.1177/0047281620901484

Allen, J. (1989). The question isn't "could" but "should": The case against using fiction in the introductory technical writing class. *The Technical Writing Teacher, 16*(3), 210–219.

Barton, B. F., & Barton, M. S. (1988). Narration in technical communication. *Journal of Business and Technical Communication, 2*(1), 36–48. https://doi.org/10.1177/105065198800200103

Blyler, N. (1998). Taking a political turn: The critical perspective and research in professional communication. *Technical Communication Quarterly, 7*(1), 33–52, doi: 10.1080/10572259809364616

Blyler, N. R. (1995). Pedagogy and social action: A role for narrative in professional communication. *Journal of Business and Technical Communication, 9*(3), 289–320. https://doi.org/10.1177/1050651995009003002

Blyler, N. R., & Thralls, C. (Eds.). (1993). *Professional communication: The social perspective.* Sage.

Bridgeford, T. (2002). *Narrative ways of knowing: Reimagining technical communication instruction* [Doctoral dissertation]. Michigan Technological University.

Bridgeford, T. (2004). Story time: Teaching technical communication as a narrative way of knowing. In T. Bridgeford, K. S. Kitalong, & D. Selfe (Eds.), *Innovative approaches to teaching technical communication* (pp. 111–134). Utah State University Press.

Bridgeford, T. (2007). Communities of practice: The shop floor of human capital. In C. L. Selfe (Ed.), *Resources in Technical Communication: Outcomes and Approaches* (pp. 165–182). Routledge.

Bridgeford, T. (2019). Imagination as agency. In S. Flanagan, & M. Albers (Eds.). *Editing in the Modern Classroom* (pp. 66–90). Routledge.

Bruner, J. (1991). The narrative construction of reality. *Critical Inquiry, 18*(1), 1–21. https://doi.org/10.1086/448619

Colton, J. S., & Holmes, S. (2018). A social justice theory of active equality for technical communication. *Journal of Technical Writing and Communication, 48*(1), 4–30. https://doi.org/10.1177/0047281616647803

de Certeau, M. (1984). *The practice of everyday life.* (Trans. Steven Rendall). University of California Press.

Dombrowski, P. M. (1999). *Ethics in technical communication.* New York: Pearson.

Dragga, S. (1999). A question of ethics: Lessons from technical communicators on the job. *Technical Communication Quarterly, 6*(2), 161–178. https://doi.org/10.1207/s15427625tcq0602_3

Haas, A. M. (2012). Race, rhetoric, and technology: A case study of decolonial technical communication theory, methodology, and pedagogy. *Journal of Business and Technical Communication, 26*(3): 277–310. https://doi.org/10.1177/1050651912439539

Jones, N. N. (2016a). The technical communicator as advocate: Integrating a social justice approach in technical communication. *Journal of Technical Writing and Communication, 46*(3) 342–361. https://doi.org/10.1177/0047281616639472

Jones, N. N. (2016b). Narrative inquiry in human-centered design: Examining silence and voice to promote social justice in design scenarios. *Journal of Technical Writing and Communication, 46*(4), 471–492. https://doi.org/10.1177/0047281616653489

Jones, N. N., Moore, K. R., & Walton, R. (2016). Disrupting the past to disrupt the future: An antenarrative of technical communication. *Technical Communication Quarterly, 25*(4), 1–19. http://dx.doi.org/10.1080/10572252.2016.1224655

Jones, N. N., Savage, G., &, Yu, H. (2014). Tracking our progress: Diversity in technical and professional communication programs. *Programmatic Perspectives, 6*(1), 132–152.

Jones, N. N., & Walton, R. (2018). Using narratives to foster critical thinking about diversity and social justice. In A. Haas, & M. Eble (Eds.), *Key theoretical frameworks for teaching technical communication in the 21st century* (pp. 241–267). Utah State University Press.

Journet, D. (1999). The limits of narrative in the construction of scientific knowledge: George Gaylord Simpson's *The dechronization of Sam Magruder*. In J. M. Perkins & N. Blyler (Eds.). (1999). *Narrative and professional communication* (pp. 93–106). Ablex.

Kimball, M. (2006). Cars, culture, and tactical technical communication. *Technical Communication Quarterly, 15*(1), 67–86.

Loftin, K., & Cryer, M. (2015). Everyday ethics and the technical communicator. *Intercom, 62*(5), 25–26.

Longo, B. (1998). An approach for applying cultural study theory to technical writing research. *Technical Communication Quarterly, 7*(1), 153–173.

Miller, C. R. (1979). A humanistic rationale for technical writing. *College English, 40*(6), 610–617.

Moore, K. (2013). Exposing hidden relations: Storytelling, pedagogy, and the study of policy. *Journal of Technical Writing and Communication, 43*(1), 63–78. https://doi.org/10.2190/TW.43.1.d

Moore, K. (2020). *The value of technical communication in enacting social justice.* Blog post. https://www.depts.ttu.edu/english/grad_degrees/Open_Grounds/Spring_2015/moore_tcr_socialjustice.php (link is now defunct)

Orr, J. E. (1996). *Talking about machines: An ethnography of a modern job.* ILR.

Perkins, J. M., & Blyler, N. (Eds.). (1999). *Narrative and professional communication.* Ablex.

Sanders, S. R. (1985). *Terrarium.* Indiana University Press.

Savage, G., & Mattson, K. (2011a). Perceptions of Racial and Ethnic Diversity in Technical Communication Programs. *Programmatic Perspectives, 3*(1), 5–57.

Savage G., & Matveeva, N. (2011b). Toward racial and ethnic diversity in technical communication programs: A study of technical communication in historically Black colleges and universities and tribal colleges and universities in the United States. *Programmatic Perspectives, 3*(1), 58–85.

Scott, J. B., Longo, B., & Wills, K. V. (Eds.). (2007). *Critical power tools: Technical communication and cultural studies.* State University of New York Press.

Smart, G. (1999) Storytelling in a central bank: The role of narrative in the creation and use of specialized economic knowledge. *Journal of Business and Technical Communication 13*(3), 249–273. https://doi.org/10.1177/1050651999013003002

Walton, R., Moore, K. R., & Jones, N. N. (2019). *Technical communication after the social justice turn: Building coalitions for action.* Routledge.

Weaver, W. (1990). *A gravestone made of wheat. Minneapolis.* Greywolf.

Wenger, Etienne. (1998). *Communities of practice: Learning, meaning, and identity.* Cambridge University Press.

Chapter Seventeen

Finding Agency Through Tactical Technical Communication

Privacy and Data Surveillance

Sarah Young and Jason Pridmore

Concerns about the intrusive practices of surveillance have long been discussed, but more recent examples of governmental and consumer practices have made concerns more prescient. From the Snowden revelations, to fears that data gathered by period tracking apps and mobile data locations could potentially be subpoenaed and used as evidence in prosecuting women for abortions after the repeal of *Roe v. Wade* (Torchinsky, 2022), to Covid-vaccine-microchipping conspiracy theories, the desire to protect oneself from ever-invasive forms of government, workplace, and consumer surveillance have become staple news items and part of everyday conversations. Inevitably, such concerns translate into desires to protect and respond to ever-encroaching invasions of privacy and exploitations of personal data. But how are citizens encouraged to resist surveillance practices? Some suggestions are found through internet-based resources such as YouTube that offer tools, tricks, strategies, and tactics to resist surveillance. Thus, some YouTube videos can be considered forms of tactical communication that encourage viewers to resist surveillance from proprietary powers. This study focuses specifically on the knowledge found within such internet-based resources and asks what these convey to the average user. Within this context, our study examines the following research question: What surveillance scenarios and solutions are represented by the YouTube videos, and what can this tell us about surveillance and tactical communication? We ultimately argue that YouTube videos are tactical spaces that provide narratives of resistance to surveillance through privacy controls, thereby offering ways to resist, even if limited in the way "resistance" is framed, and the videos ask us to rethink who proprietary powers are in the first place that tactics work around.

Tactical Communication, Surveillance, Privacy, Resistance, and Agency

Tactical communication can be a way people share knowledge they have learned either inside or outside of organizational cultures to others in informal ways (Kimball, 2006). Tactical methods work around institutional strategies and official forms of communication. Video-based tactical communication has proliferated on the video-streaming platform YouTube. Producers of all types can post content there, and one particular genre of YouTube videos that are examples of tactical communication are videos describing antisurveillance (often pro-privacy) techniques for the protection of bodies and data on technological devices and elsewhere. This genre of video is particularly tactical because not only is the YouTube platform facilitating a tactical means of delivery for the producer, giving suggestions to work around those in power, but the content of the videos is also tactical because the viewer learns beyond institutional confines.

Surveillance and resistance are both closely linked to the idea of tactics, especially through the concept of institutional power. Surveillance is often linked to power and often requires the authority or ability to watch over someone else, and those that have the power to watch over others are often institutions. To resist surveillance then, also means resisting an institution, which is also what tactics are: resistance to institutions. Even de Certeau (1984) mentions that to engage in tactics, one must avoid "the surveillance of the proprietary powers" (p. 73). For surveillance, some acts of resistance range from more artistic acts like staging fake Google Street View scenes (Ingraham & Rowland, 2016) to more destructive practices like breaking surveillance devices (Marx, 2003) to manipulating other forms of technology or being an informed consumer. Cross-cutting the recommendations, often such as the case for this study, anti/counter-surveillance tactics advocate for increased privacy.

To these ends, to answer our driving question—What surveillance scenarios are represented by the YouTube videos, and what can this tell us about surveillance and tactical communication?—we explore the following questions: Where do YouTube videos identify instances of surveillance? What tactics are content creators on YouTube videos advocating? What spaces of agency are opened up?

Methods

This study was conducted following a thematic analysis approach. Braun and Clarke (2006) describe a theme as a "pattern" (p. 80) and thematic analysis as identifying, analyzing, and reporting these patterns in the data (p. 79). Exploring and recording these themes help to describe a larger topic and can utilize keywords or phrases that indicate similar patterns of data. The process involves searching across different types of documents to find repeated patterns, as well as moving back and forth between writing and revising codes. One important part of thematic analysis is that researchers focus on particular aspects in the data as part of the choices that they are making in their analysis process. As such, identifying themes is a matter of both analysis and interpretation, and themes don't just "emerge" but are rather a mixture of both the data and one explanation of the data in relation to the topic.

After selecting ten videos (see Young, 2022) that overlapped three search phrases (*protecting my data*, *avoid digital spying*, and *avoiding data surveillance*) on three different browsers (Chrome, Firefox, and Edge), we analyzed the videos to determine what surveillance scenarios and tactics were being represented. To do this, we chose six aspects to identify: (1) act, (2) target, (3) agent, (4) site, (5) consequences of the acts, and (6) instructions recommended to the audience (the suggested tactics). These six aspects were chosen for analysis because they represented the surveillance scenario being represented during the videos. Some videos focused on several scenarios, and others focused on just one. To define the terms a bit further, an *act* is the act being committed in which someone would need to avoid or protect themselves from. The *target* is who is the target of that act. The *agent* is the one carrying out that act, and the *site* is where the surveillance act was said to occur. The *consequences* are what results from the act, and the *what-to-do tactics* are the suggestions that the creators mentioned or the tactics to get around the acts. The criteria for the surveillance scenario were inspired by the categories used in the privacy work of Mulligan and colleagues (2016) as well as whistleblowing research by Jubb (1999). Mulligan and colleagues (2016) proposed an analytical tool for mapping contested spaces of privacy in various contexts with 14 dimensions, such as object, target, subject, action, offender, justification (p. 10). Jubb (1999) uses six elements like a disclosure's act, subject, and recipient, as well as actors, targets, and outcome to aid in determining cases of whistleblowing (p. 77). These aspects also sound like the work of Burke's (1945) dramatistic pentad with the rhetorical elements of act, scene, agent, agency, and purpose, but while the terms and some of meanings of these aspects can overlap with Burke, the explanation of these elements is drawn more directly from a mix between Jubb (1999) and Mulligan and colleagues (2016).

Results

With that framework, we identified various patterns for each aspect of act, target, agent, site, consequences, and what-to-do tactical recommendations.

Act

The notion of *act* is action-oriented, and defining an act of surveillance is connected to identifying the verb through which it was carried out. The various verbs involved in the compromise are listed in table 17.1. The categories for act resulted in the actions taken that could compromise at least five elements: (1) location, (2) body, (3) data, (4) technology, and (5) passwords. For each category, it is useful to clarify that the category boundaries are not rigid, and some elements could fit various categories. Particular activities were sorted depending on the perceived emphasis of the act. For instance, identify is in both location and body, but identify location is trying to determine where one is, and identify body is to determine who one is.

Going through each of the categories a little further, if one's *location* was compromised, that means someone was able to find or monitor one's location. For the *body* category, one's body (focusing on the body rather than the location) would be a target of the act in some form, such as using one's body to forcefully open biometric passcodes. If the activity was classified as *data*, then somehow the data would be the focus of the act, such as hacking it or tracking it. If tech was the category, then something

Table 17.1. Results and Examples of Act. The table provides sample verbs for each act of surveillance. The *act* is the activity being committed by the actor from which someone would need to protect themselves.

Act	Example of Act
Location	Find Geotag Identify Monitor Predict Reveal Track Watch
Body	Assess Document Defraud Identify Impersonate Intercept Link List Listen Monitor Recognize Scan Shadow Spy Surveil Target Track Use Watch
Data	Access Analyze Breach Collect Drag Detect Exfiltrate Fake Follow

continued on next page

Table 17.1 Continued.

Act	**Example of Act**
Data (continued)	Hack Intercept Link Look at Move Observe Predict Record Read Scan Seize Steal Store Subjected Surveil Target Theft Track Use
Tech	Compromise Exploit Hack Invade Intercept Monitor
Password	Access Avoid Break Cycle through Find Hack Use

Source: Created by the author.

was going to happen to a piece of technology, ranging from a general idea of compromise to a lengthier monitoring of a device. Finally, if a *password* was being targeted, then one's password was the way in which a compromise would come about. It is of note that the password code could indeed fall under the tech category, but again, the emphasis occurred at enough frequency that password became its own focus.

Target

Targets were the entities who were recipients of the act by the actor. Table 17.2 provides a general overview of this information.

In alphabetical order, we placed some entries into a *bad actor* category because it was noted that some surveillance targets those involved in criminal activity. This category was not the main focus of many videos, however, and was often mentioned to show that surveillance should target these individuals but often extends to *you* or other targets. Examples of bad actors include abusers or military targets, suspects, murders, pedophiles, ISIS, and Al Qaeda. This category was a departure from the other categories in that the surveillance for these targets was talked about in a positive manner. Surveillance against other targets was discussed negatively. Cities and communities were also mentioned in a more collective idea that particular areas or communities might be targeted, either with a generic idea of cities, civilians, or citizens or more specific groups such as Chinese citizens and low-income African Americans. *Companies* could refer to generic or particular

Table 17.2. Results and Examples of Target. The *target* is the recipient of the listed act.

Target	Example of Target
Bad actors	Individuals who commit crimes or abusive individuals.
Cities/ communities	Cities in general or particular communities such as low-income African American or Muslim communities.
Company	An employer, with the focus on the business rather than the person.
Others	Generic "others" or miscellaneous, such as consumers or the abused.
Protest(ers)/ activists	Protests and protesters/those standing up for others.
Us	A collective group.
Users	Users of a particular technology.
You	Addressed to "you" and your technology or data in the second-person perspective.

Source: Created by the author.

corporations. Another target was the *other*, which is a collection of others that either were not specifically identified or a more miscellaneous group of entities that didn't fit into other categories such as wives, sources, or shadows. *Protestors* and *activists* like those standing up for abuse were also signaled out as particularly threatened by surveillance practices. The agents in this case could even be police who were uncomfortable with protests. There was also a more inclusive idea of *us* that in a general way drew the audience into a collective. For instance, we as a collective us can be under the watch of police surveillance. The category of technology users was also referenced, like users of Google can be tracked by Google. Last but not least, one of the largest categories of targets was *you* with a second-person perspective in that you the viewer could be susceptible to any of these acts. Your body can be watched. Your data could be stolen. You play a role in preventing and avoiding surveillance.

Agent

Agents were the ones carrying out the act of surveillance, and these agents could be sorted into four categories: governments, corporations, people, and technology. Table 17.3 provides a general overview of this information.

The first category of agent was the *corporation*. Videos stressed that corporations like Google or Facebook watch and track one's data. Second, the *government* category resembled the corporation, but instead of a company watching data, some part of the government was the agent carrying it out. Third, the code of *people* was used whenever a particular human was described as carrying the act of surveillance, like abusers, bad actors, thieves,

Table 17.3. Results and Examples of Agent. The *agent* is the one carrying out the listed act.

Agent	**Example of Agent**
Corporations	A corporation such as Google or Facebook is the operative.
Government	Government entities are responsible for the act of surveillance.
People	A category of person is the agent, be it a thief, jealous husband, unnamed other, et cetera.
Technology	Technology is the focus (rather than the human part of the device), such as an app or gray key box.

Source: Created by the author.

unauthorized parties, insiders, hackers, and attackers. The people that carried out surveillance were most often out to do harm rather than use surveillance for positive reasons. Finally, *technologies* themselves were coded as agents when the technology was described as a more deterministic agent carrying out the act (such as invasive apps or technology). Examples included gray key boxes that can cycle through passwords or other types of surveillance-capable technologies like predictive policing software or stingrays. It is useful to clarify, too, that for the government, corporations, and technology codes, the emphasis was on the agency or technology rather than the officers or humans that work with the technology or for the companies or government.

Overall, it is interesting to note that there was mostly a negative connotation to surveillance, and although some positive aspects of it were discussed such as targeting criminals, surveillance was largely positioned as a more negative event, with the agents being abusers, thieves, and bad guys, secretive government entities or police using military-grade technologies on their citizens, or deterministic technologies. Agents in these surveillance scenarios were those to be avoided.

Site

Sites ranged from in-person to virtual spaces to mobile/phones. Table 17.4 illustrates samples of the categories.

The *in-person* site was a more physical location of surveillance, and examples are the physical body being shadowed by an operative, unshredded documents that thieves can steal from the trash, or police watching the physical location of citizens with drones. Surveillance on *the internet and online sites* were the virtual or online acts one might be subjected to, and these ranged from anywhere virtual where your data is stored, such as the databases operated by companies, or attackers may give you fake links

Table 17.4. Results and Examples of Site. The *site* is where the surveillance act was said to occur.

Site	Example of Site
In-person	The act involves the physical, material, and in-person world.
Internet/online	The act takes place online, digitally, or virtually.
Mobile/phone	The act takes place primarily on a phone.

Source: Created by the author.

to click on where they can then steal your personal information. Finally, the *mobile/phone* category showed how one's mobile phone can facilitate all the listed acts of surveillance from revealing a location, identifying a body, allowing data to be tracked, exploitation of technology through vulnerabilities like nonupdated apps, and risk to passwords. Phones weren't limited to mobile sites, though; risk could also come from phishers trying to get information via a landline phone, or attackers may give your false phone numbers to impersonate institutions, hoping you will trust and call them to provide personal information. It is of note that one can access virtual locations via one's phone, so borders in this category can be porous, but some situations privilege one category over the other such as when emphasis is on the phone's connectivity rather than the internet in general.

Consequences

Consequences were a bit tricky to identify because it was harder to differentiate between act and consequence as what consisted of both was more porous and mutable. Consequences are what results from the act, but sometimes, the act could almost be the consequence. For instance, some videos described that an act was the action of hacking into someone's data, which could lead to the consequence of identity fraud, but some videos either ended at the fact that hacking was the consequence or also described hacked data as a consequence of the act of breaking a password. With the complexity, we did our best to categorize the consequences coming back to the definition that consequences were the results of the act. We then put these results in eight general ways as detailed in Table 17.5.

First, there were consequences to the *body* as in one's physical body, such as when your body is forcibly used to activate biometric passwords. A second consequence was a compromise of *communications*. Your communications can also be compromised because Google can read your emails or stingrays can intercept your calls, allowing the police to monitor them even if you're not a suspect. Third, one's *data* could be compromised such as when it is stolen or accessed without authorization Fourth, *financial consequences* result when your identity is stolen, or your company can be liable for stolen data. Fifth, the generic *other* category represented more one-off situations, such as the case of one video that described how to engage in counter-surveillance and "out" a shadow. The internet is more efficiently censored, or one faces consequences for speaking out. Sixth, the target could

Table 17.5. Results and Examples of Consequences. The *consequences* result from the act.

Consequences	Example of Consequences
Bodily harm	One's physical body is compromised in some way.
Communication	Communications such as messages or emails are compromised.
Data	A target's data is affected, corresponding to the act carried out.
Finances	Your finances are at risk.
Generic/other	This miscellaneous category consists of items that don't fit into other categories.
Identified	A target is identified in some way.
Location/Located	A target is either located, or their location is compromised.
Tracked/predicted behavior	One is tracked virtually or in person, and future behaviors can be predicted as a result.

Source: Created by the author.

be susceptible to being identified *in name*. If one's name is tied to their face or other data, others can figure out who they are, or the police can list them negatively as a gang member or activist. Seventh, one could be vulnerable to location identification as to where one is and where one moves. Finally, if one's location or data is tracked, this could lead to *prediction of future* behaviors like the probability of product purchases.

What-to-Do Tactics

Finally, we categorized recommendations in thirteen ways, ranging from paying attention and adapting privacy controls for one's apps, browsers, cloud, connectivity, emails, messenger, passwords, and phone. It was also advised to adopt a new mindset altogether, visit other work by the content provider, and monitor accounts, protect one's personal information, and a final category encompassed consequences that did not really fit a larger pattern. Table 17.6 provides a summary of the codes and an explanation of their classifications.

Table 17.6. Results and Examples of What-to-Do Tactics. The *what-to-do tactics* are the suggestions that the creators mentioned or the tactics to get around the acts.

What-to-Do Tactics	**Examples of What-to-Do Tactics**
Apps	Apps can be a source of concern for things like locations and data privacy but can also be used to help be more secure.
Browser	Creators argue that not all browsers are designed for privacy, and users should be choosy about what services they use.
Cloud	Recommendations range from not using the cloud to using the cloud more safely or extensively.
Connectivity	There are ways to keep the connections to the internet more secure.
Content creator	Watch additional content from the creator for more information.
Email	The audience is encouraged to pay attention to their email and email services.
Message	Pay attention to messengers and use tools such as encryption.
Mindset	Being more aware of privacy and surveillance requires a mindset and sensitivity toward the topics.
Monitor accounts in general	Take general care of your accounts (monitor your credit, etc.).
Passwords	Keep in mind that passwords are both vulnerabilities as well as spaces to help maintain privacy.
Personal information	These recommendations focus on limiting the distribution of personal identifiable information.
Phone	Suggestions revolve around working with the settings of one's phone.
Other	This miscellaneous category consists of tips that don't fit into broader categories, such as shredding paperwork or updating devices in general.

Source: Created by the author.

Examples to illustrate the codes are that first, for *apps*, suggestions are to use progressive web apps, and not to let apps refresh in the background to minimize location sharing. Suggestions for browsers are actions such as using privacy-respecting alternative browsers or to make sure you use secure sites when shopping. To avoid problems with the *cloud*, the videos suggested to not use your iCloud in the first place or use Google's cloud services. *Connectivity* examples were using VPNs to control connections or avoid open networks for activities such as looking up finances. One slightly different recommendation was to follow the *content creator* for more tips or to watch other surveillance videos from the account. To keep *email* more secure, it was advised to use encryption, use separate emails for different utilities to minimize online tracking, and minimize giving out personal information via email. To keep *messages* more secure, viewers were encouraged to encrypt messages (which also relates to using apps). A *mindset* was also important because it creates useful practices like compartmentalization by minimizing connecting accounts and encourages companies that store sensitive data to be prepared for data breaches. Another tactic is to *monitor accounts*, such as one's physical mail and watch credit card statements, also making sure your technology accounts are up to date. To avoid problems with *passwords*, it was recommended to use a password manager, don't remember passwords on your computer, and don't use the same password for multiple accounts. It was also advised to avoid distribution and protect *personal information* by doing things such as restricting access to social security numbers and not putting personal information on social-media sites. The *phone* was an often-recommended tactic, to use such as taking advantage of the iPhone's built-in measures to deter hacking or using your phone to assist in two-factor authentication. Finally, the *other* category was for the other suggestions that didn't have a definitive pattern, such as don't use Google Maps or Siri, don't search with YouTube, and make sure to shred all important documents.

Implications

A look at these results illustrates that indeed, content creators are delivering content that provides tactics to resist surveillance. The creators identify a range of acts that can be considered surveillance, a range of actors that carry out surveillance, a range of targets, a range of consequences, and provide a range of tactics one can perform to resist this surveillance. This is useful for anyone looking to understand how surveillance is carried out, how potential targets are taught to resist through tactics, and how tactical communication on YouTube aids in this instruction. The recommendations, however, are also illuminating in two other ways: one that is focused more on surveillance and one that is focused more on tactical communication.

For the former focus, these recommendations highlight a limited concept of surveillance resistance previously noted in surveillance literature. Although the videos do provide tactics that provide some form of agency for the targets, the recommendations focus mostly on the individual and fail to encourage more collective ability to take action. Most of the videos focused on how an individual can minimize their own data leakage, and this reinforces the individualized approach that encourages personal privacy fitness rather than collective change. As Fernandez and Huey (2009) note, themes of resistance around personal protection can superficially make people feel better but actually imply that "protection" from the "surveillance threat" is just "a series of measures undertaken by individuals, hiding the collective possibility for resistance" (p. 198). The disproportionate relationship in which your only response to a much larger and pervasive corporation is simply about making sure your privacy settings are up to date can perpetuate a feeling of helplessness, and this small bandage approach to systemic problems feeds into postprivacy mindsets (Burkart & Andersson Schwarz, 2014) where some individuals may cease to care about privacy and surveillance due to the enormity of the system and the triviality of the control. We argue that while personal protections are useful, collective ideas of rights to information can also empower users to be more conscious of their privacy. More research would be useful in exploring more collective calls for resistance to say advocate for the change of corporate policies of consumer data collection, in comparison to the individual privacy hygiene narrative in YouTube videos, especially seeing if particular content creators might call on collective rights to privacy more often.

The second critique of the results centers on tactical communication. When looking at the YouTube videos, especially scenarios involving actors such as hackers/attackers, the tactics presented by the creators react to the hackers, not what might be considered strategic, proprietary powers. Sometimes the tactics even call for the renewed attention to the measures set up by the "powerful" or proprietary corporations. For instance, one repeated tactic was to revisit the factory settings for one's phone to make sure the user's setting matched their objectives. Thus, this area then calls attention to what is meant by *proprietary power* because the tactics here react to the hackers who themselves reacted to the more formal institution. This then begs the question, Is the communication tactical if it does not react to a formal, strategic institution? Or can/does power transfer to other entities, such as where a hacker might be considered a strategic proprietary entity that thus, "you" may then tactically resist, thus possibly even resisting by resorting back to the strategic design of the propriety power? In other words, what qualifies as the strategic, proprietary power, who are the "weak," and do these positions transfer? Are there layers of who is or can be considered tactical or strategic? In the case of this study, in some instances, hackers reacted to the strategies, and then audiences were told to resist the hackers, producing several layers of strategic and tactical.

Conclusion

Overall, this study provides a window into what knowledge videos provide an audience when searching for ways to avoid surveillance. In terms of act, agent, target, site, consequences, and what-to-to tactics, a wide variety of scenarios were presented. Future research could examine the conclusions drawn here in a larger scale, but this research provides a useful snapshot into seeing how audiences are taught they are surveilled as well as taught how they can resist.

Deeper implications are twofold. For the agency to resist surveillance, content creators usefully publish tactics of surveillance, often along the lines of increased privacy controls. While useful, these are limited and could benefit from more collective calls for privacy as well as more tactics that an individual could employ that aren't focused specifically on privacy.

These conclusions are useful not just for content creators making those videos but also for researchers who look at what surveillance and privacy narratives are circulating in internet cultures. Further, for tactical communication, there is a question of reconsidering what it means to be tactical to either bear down and consider strategies as only the official strategies that large institutions might offer or open up and to consider that tactical communication might also be responding to questionably strategic actors such as hackers or other "bad guys." Overall, both results help illustrate the importance of looking at tactical spaces that are created and seen by many viewers across geographically dispersed places, contributing to internet culture of narratives of choice, agency, and resistance and informing a conversation about what is posted and consumed in tactical, extra-institutional spaces.

References

Burkart, P. & Andersson Schwarz, J. (2014). Post-privacy and ideology. In André Jansson & Miyase Christensen (Eds.), *Media, surveillance and identity* (pp. 218–237). Peter Lang.

Burke, K. (1945). *A grammar of motives*. University of California Press.

Braun, V. & Clarke, V. (2006). Using thematic analysis in psychology. *Qualitative Research in Psychology, 3*(2), 77–101. https://doi.org/10.1191/1478088706qp063oa

de Certeau, M. (1984). *The practice of everyday life* (S. Rendall, Trans.). University of California Press.

Fernandez, L. A. & Huey, L. (2009). Is resistance futile? Some thoughts on resisting surveillance. *Surveillance & Society, 6*(3), 198–202. https://doi.org/10.24908/ss.v6i3.3280

Ingraham, C., & Rowland, A. (2016). Performing imperceptibility: Google Street View and the tableau vivant. *Surveillance & Society, 14*(2), 211–226. https://doi.org/10.24908/ss.v14i2.6013

Jubb, P. B. (1999). Whistleblowing: A restrictive definition and interpretation. *Journal of Business Ethics, 21*(1), 77–94.

Kimball, M. (2006). Cars, culture, and tactical technical communication. *Technical Communication Quarterly, 15*(1), 67–86. https://doi.org/10.1207/s15427625tcq1501_6

Kimball, M. (2017). The golden age of technical communication. *Journal of Technical Writing and Communication, 47*(3), 330–358. https://doi.org/10.1177/0047281616641927

Marx, G. (2003). A Tack in the Shoe: Neutralizing and resisting the new surveillance. *Journal of Social Issues, 59*(2), 369–390. https://doi.org/10.1111/1540-4560.00069

Mulligan, D. K., Koopman, C., Doty, N. (2016). Privacy is an essentially contested concept: a multi-dimensional analytic for mapping privacy. *Philosophical Transactions A.*, 374. https://doi.org/10.1098/rsta.2016.0118

Torchinsky, R. (2022, May 10). *How period tracking apps and data privacy fit into a post-*Roe v. Wade *climate*. NPR. https://www.npr.org/2022/05/10/1097482967/roe-v-wade-supreme-court-abortion-period-apps

Young, S. (2022, July 23). *Finding agency through tactical technical communication: Included videos*. https://sites.google.com/view/sarahyoungphd/foundations-of-research/surveillance-technical-communication/finding-agency-through-tactical-technical-communication?authuser=0

Contributors

Sandy Brack is a PhD candidate at the University of Louisiana-Lafayette. She is particularly interested in deconstruction as a theoretical framework to study onto-epistemologies in postmodern American literature and contemporary discourse formation. Her research has explored areas including Latin American decolonization, eco-feminism, Afrofuturism, and eco-science fiction. She is completing her PhD as a University of Louisiana Fellow.

Tracy Bridgeford is a professor of technical communication at the University of Nebraska at Omaha and editor for *Technical Communication Quarterly*. Her primary research focus has been on technical communication, historical and current pedagogy, and practice. She edited *Teaching Professional and Technical Communication: A Practicum in a Book* (University Press of Colorado/Utah State University Press) and edited/coedited three collections as part of the Association of Teachers of Technical Writing Series in Technical Communication (Taylor & Francis/Routledge), including *Teaching Content Management in Technical and Professional Communication, Academy-Industry Relationships*, and *Sharing Our Intellectual Traces.*

Ryan Cheek is an assistant professor of technical communication at Missouri University of Science and Technology. His work has been featured in impactful academic field publications such as *Technical Communication Quarterly, Journal of Technical Writing and Communication*, and *Communication Design Quarterly*. Driven by a fascination with how individuals, societies, and institutions navigate the complicated intersection of communication ethics and political technology, Dr. Cheek's research leverages rhetorical methods to investigate the role of technical communication in shaping ideologies, technologies, and eschatologies within democratic politics.

Sara C. Doan (she/her) is an assistant professor of experience architecture in the Writing, Rhetoric, and Cultures Department at Michigan State University. She examines how expertise is framed and enacted across different genres, such as instructor feedback, misleading data visualizations about COVID-19, vaccine hesitancy, and the content strategy of southeastern state health departments. Her work on data visualizations, preventive health behaviors, and feedback in technical communication courses has appeared in the *Journal of Business and Technical Communication*, *IEEE Transactions on Professional Communication*, and *Business and Professional Communication Quarterly*.

Avery Edenfield, associate professor of rhetoric and technical communication at Utah State University, works at the intersections of technical, cultural, and public rhetorics, with attention to the technical-writing strategies marginalized communities employ for self-advocacy, particularly in extra-institutional contexts. His work has appeared in the *Journal of Business and Technical Communication*, *Technical Communication*, *Nonprofit Quarterly*, *Technical Communication Quarterly*, and *Communication Design Quarterly*.

Brian Fitzpatrick is an associate professor at George Mason University, where he teaches composition and literature. His research is primarily focused on workplace writing, as well as online and hybrid learning. He is the cofounder of the Archive of Workplace Writing Experiences and was recipient of the Conference on College Composition and Communication's Emergent Researcher Award for 2017 through 2018 and a CCCC Research Initiative Grant for 2021 through 2022. His work has appeared in *WPA Journal*, *Performance Improvement Quarterly*, *Academic Labor: Research and Artistry*, and *Double Helix*.

Guiseppe Getto is an associate professor of technical communication and director of the MS in technical communication management at Mercer University. His research focuses on utilizing user-experience (UX) design, content strategy, and other participatory research methods to help people improve their communities and organizations. He has published two books: *Content Strategy in Technical Communication* and *Content Strategy: A How-to Guide* (both with Routledge). He has also published numerous articles in journals such *as IEEE Transactions on Professional Communication*; *Technical Communication*; *Computers and Composition*; *Rhetoric, Professional Communication, and Globalization*; and *Communication Design Quarterly*. Visit him online at: http://guiseppegetto.com.

Steve Holmes has been an associate professor at Texas Tech University since 2019. He has published broadly in digital rhetoric, writing studies, and technical communication, with a specific focus on ethics and technology. He has authored two books: *Procedural Habits* (2017) and *Rhetoric and Technology and the Virtues* (2018) with Jared S. Colton. Most recently, he coedited *Reprogrammable Rhetoric: Critical Making Theories and Methods in Rhetoric and Composition* (2022) with Michael J. Faris.

Emily January, associate professor of English at Weber State University, has conducted qualitative research in the United States, India, South Africa, and Botswana. Her research focuses on professional identities and organizations from a feminist perspective by examining social media, analyzing archival sources, and conducting interviews. Her work has appeared, among other places, in *Technical Communication Quarterly*, the *Journal of Business and Technical Communication*, and *Peitho*.

Michael Knievel is an associate professor at the University of Wyoming, where he teaches courses in composition and professional writing and directs the first-year writing program. His research interests include police-related writing and rhetoric, the relationship between technology and the humanities, and small technical-communication programs. His work has appeared in publications such as the *Journal of Business and Technical Communication*, *Computers and Composition*, *Programmatic Perspectives*, and the *Journal of Technical Communication and Writing*, as well as several edited collections.

Brad Lucas is an associate professor of English at Texas Christian University, where he has served as department chair and director of both undergraduate and graduate programs. He is the former editor of *Composition Studies* and the author of *Radicals, Rhetoric, and the War: The University of Nevada in the Wake of Kent State*. In addition to studying technical-writing curriculum and pedagogy, he is investigating the extremist rhetorics of sixties revolutionary groups and their turn to militancy.

Jessica McCaughey, PhD, is an associate professor in the University Writing Program at George Washington University, where she teaches academic and professional writing. Jessica's research focuses primarily on the transfer of writing skills in professional settings. Along with her colleague Brian Fitzpatrick, she cofounded and codirects the Archive of Workplace Writing Experiences.

Dannea Nelson is a PhD candidate at the University of Otago. She holds an MA in English, a BS in history, and a BA in English from Weber State University, where the research for this contribution occurred. Dannea has presented her work at the National Conference on College Composition and Communication and the inaugural Utah Southwest Regional Conference on Student Research in Gender and Women's Studies conference held at Brigham Young University in 2022. Dannea's research interests lie in the Gothic, crime writing, LGBTQ+ expressions of gender and sexuality, feminist and queer rhetoric in popular culture, and social justice and interdisciplinary studies.

Jason Pridmore is the vice dean of education for the Erasmus School of History, Culture, and Communication and an associate professor in the Department of Media and Communication at Erasmus University Rotterdam. Jason coordinates both SEISMEC, an EU-funded project that focuses on piloting Human Centric Industry innovations across multiple national and industrial contexts, and COALESCE, an EU-funded project focused on developing the European Competence Centre for Science Communication. His primary research focus is on practices of digital identification and monitoring, digital science communication, the use of new/social media and consumer data as surveillance, and digital (cyber) security issues.

Shannon N. Sarantakos is an assistant professor at Farmingdale State College. Her research interests center on visual rhetoric and data visualization in professional- and technical-communication settings. Dr. Sarantakos is particularly interested in the rhetorical complexity of scientific images and displays, especially as it contributes to communicating complex health information. She also writes about academic motherhood. Sarantakos's work has been featured in *Journal of Technical Writing and Communication*, *Technical Communicator Quarterly*, *Kairos*, and *Journal of Mother Studies*.

John L. Seabloom-Dunne is the visiting assistant professor of technical and scientific writing at the Ohio State University's Department of English. He studies the practices and histories of technical communication. Some of his current areas of interest include DIY cultures of instruction, speculative visions of technology, and the friction that emerges from familiar genres in unfamiliar contexts.

John T. Sherrill is an independent scholar who completed his degree in English—Rhetoric and Composition from Purdue University in 2019. He returned to the United States after four years as an assistant professor of professional writing at Qatar University in Doha, Qatar. His research focuses on intersections among feminism, technical communication, automation and AI, and DIY/Craft/Hacker/"Maker" communities. His work has been published in *Communication Design Quarterly*, *Kairos*, and the proceedings of ACM's SIGDOC.

Walker P. Smith (he/him) is assistant professor of rhetoric and composition at Goucher College in Baltimore, Maryland, where he teaches courses in academic writing, professional and technical communication, and queer rhetorics. His research explores critical theories and histories of rhetoric, sexuality, and religion across technical documents, archival ephemera, and films. His work appears in *Peitho*, *Unsettling Archival Research (SIUP)*, and *Across the Disciplines.*

Don Unger (he/him/his) is an assistant professor in the Department of Writing and Rhetoric at the University of Mississippi. His scholarship focuses on community writing and publishing and includes serving as the managing editor of Spark: A 4C4Equality Journal. His research has been published in *College English*, *Computers and Composition*, *Constellations*, *Community Literacy Journal*, *Journal of Multimodal Rhetorics*, *Peitho*, *Comics Grid*, and various edited collections.

Kevin Van Winkle, PhD, serves as an associate professor of technical communication and rhetoric at Colorado State University Pueblo. His research focuses on visual rhetoric, tactical technical communication, and workplace process improvement. His work has been featured in *Technical Communication Quarterly* and *Communication Design Quarterly.*

Joseph E. Williams is assistant professor at Louisiana Tech University. His diverse research interests include visual rhetoric, tactical technical communication, risk communication, usability, intercultural communication, and often, a synthesis of these subgenres. His work has been featured in *Journal of Technical Writing & Communication, Soil & Tillage Research*, and *Construction & Building Materials*, among others. Currently, he is working on an analysis of digital immersive exhibits.

Sarah Young is a postdoctoral researcher in digitalization and AI with the department of Media and Communication at Erasmus University Rotterdam. She does research on surveillance, privacy, communication, writing, and technology's social impact. Her current work explores the technical discourse of quantum technologies, and past projects have focused on government-surveillance rhetorics, professional communication and law enforcement's use of AI, and surveillance and privacy themes in technical communication. In addition to journal articles, she has also published *Working Through Surveillance and Technical Communication: Concepts and Connections* (SUNY Press) and coedited *Superheroes and Digital Perspectives: Super Data* (Lexington Books).

Editors

Miles A. Kimball, PhD, studies the relationships between technology and humanity. His concept of 'tactical technical communication' highlights the ubiquity of technical communication in our society. He has published broadly on e-portfolio pedagogy, information design, digital humanities, and the history of data visualization. Professor Kimball coedits (with Derek Ross and Hilary A. Sarat-St. Peter) the SUNY Technical Communication book series. The Society for Technical Communication named him an Associate Fellow in 2020 and awarded him the Ken Rainey Award for Distinguished Research in 2019.

Hayley McCullough, PhD, currently serves as the assistant professor of technical communication and digital media at New Mexico Tech. Their work explores the intersections of technology, identity, and creativity, and it examines the ways storytelling functions as a form of cultural documentation—acting as both a mirror and shaper of culture.

Hilary A. Sarat-St. Peter is an associate professor of technical communication and rhetoric at Columbia College in Chicago.

Index